"Once in a great while there appears a book that should have been written long ago. *Canine Reproduction* by Phyllis A. Holst, M.S., D.V.M. is such a book." -- *Helma Weeks, Daily Intelligencer*

"When you read it, you'll agree with me that Holst's book, *Canine Reproduction A Breeder's Guide* should never go out of print...This is a fine, fine book. There will be an empty place on a serious dog lover's library shelves if this book isn't there for consultation and review."--*Kathryn Braund, The Poodle Review*

"If you breed or are considering breeding dogs, this book is a must for your book shelf. ...Now that I have read it a second time. I find myself referring to it constantly. It has, in fact, become one of the most frequently referred to books in my library."--*Bobbie Christensen,  Hunter's Whistle/AKC*

"This is a book you are going to want (need!) on your reference shelf if you are a serious breeder."--*AJB, Dogs in Canada*

"I would like to highly recommend one particular book: *Canine Reproduction*, By Phyllis Holst, DVM, MS, published by Alpine Publishing Co. I even bought a copy for my veterinarian to have!"--*Marcia Schlehr, Golden Retriever News*

"...it is one of the most informative books available on the subject of canine reproduction. Even the experienced breeder can learn from this book."--*Cindy Williams, Editor, Newfoundland Dog Club of Canada newsletter.*

# CANINE REPRODUCTION

## A Breeder's Guide

# CANINE REPRODUCTION
## A Breeder's Guide

**Phyllis A. Holst, MS, DVM**

Foreword by Quentin LaHam, Ph.D.

Illustrated by M. Lynne Kesel, DVM

Alpine Publications
P.O. Box 7027 • Loveland, CO 80537

iv

**Credits**

Cover Photo: Bobbie Christensen
Cover Design: Joan M. Harris
Technical Photos: Phyllis A. Holst, DVM
Layout: Susan Allard
Typesetting: Hope Hicks, Artline

International Standard Book No. 0-931866-21-9

Printed in the United States of America.

*TO THE LOVE OF PUREBRED DOGS*

*How they have influenced my life!*

vi

# ACKNOWLEDGEMENTS

The original seed that finally grew into this book did not originate with me, but with the late Barbara Hagen Rieseberg and Betty Jo McKinney of Alpine Publications. They were convinced that there was a need for a serious, in-depth work on reproduction in dogs—something beyond the novice level—and their ideas and patience are deeply appreciated.

Dr. Robert Phemister, currently dean of the College of Veterinary Medicine, Colorado State University, was my first boss. More important, he was my best teacher, and the research we did together from 1966 to 1973 concerning the reproductive processes of Beagles made a real and lasting contribution to our knowledge. He taught me how to look at information with a critical eye, how to be careful in what conclusions are drawn from the information, and how to write carefully. He taught me habits that are more precious than gold. His support and assistance in reviewing the manuscript are greatly appreciated.

The photomicrographs of vaginal cells were made using the microscope and camera of the Department of Anatomy, College of Veterinary Medicine, Colorado State University. A great debt of gratitude is due to Dr. Robert Kainer for his help, ideas, and time, which made these important photos possible. Several drawings also owe their final accuracy to Dr. Kainer.

The most frightening aspect of this work was worrying that something might be stated which was inaccurate, outdated, or simply impossible to substantiate. Dr. Patricia Olson of Colorado State University, Diplomate of the American College of Theriogenology, helped ease these fears by reviewing the manuscript and setting me straight on several important points. Thank you, Patty, for your interest and support.

Several dogs served as models for photographs, and without them, many of the illustrations would not have been possible. Thanks go to the staff of the Longmont Humane Society for providing dogs and for their assistance in the photography. Also to Bobbie Christensen, Rocky Mountain Training Kennels, for her eagerness to help in obtaining photographs and for the lovely photographs of her own.

# TABLE OF CONTENTS

x

# ABOUT THE AUTHOR

The author has always thought of herself as a late bloomer when it came to finding a career for herself—that special niche which every person must find for his/her life. Her interest in purebred dogs and career in veterinary medicine developed at about the same time, several years after graduating from college with a degree in Botany and French.

Her first gainful employment after college was as a histologic technician at the Collaborative Radiological Health Laboratory, Colorado State University. The lab was, and still is, a major research center studying the long-torm effects of low-level irradiation on Beagles. Part one of the project was being undertaken at that time: to study prenatal development in the dogs so that the organism being irradiated could be better understood. The director of the lab, Dr. Robert Phemister, was especially interested in the reproductive research, and under his leadership, Phyllis became more involved in research, growing gradually away from the histopathology lab.

A master's degree came as a natural outgrowth of the research, and her master's thesis was a study of the development of the Beagle embryo from conception to implantation—work that had not been done in a complete way before. The thesis, along with several other studies, was published and is now a permanent part of the literature. The time of ovulation and its relation to LH levels and to vaginal cytology were important parts of the research being done at that time. Recognizing the onset of diestrus and its relationship to other important reproductive processes is probably the single most important knowledge to have come from the research.

Shelly came into the author's life shortly after Phyllis settled into a career. Shelly was supposed to be a Sheltie but immediately proved to be of somewhat mixed heritage. She was enough like a Sheltie, though, that it was not long before a show quality Sheltie bitch joined the family. Jolee was the foundation for Sarabande Shelties, and as a first dog could not have been more ideal. She won eight championship points and CDX under the handling of her novice owner. She provided the opportunity to become involved in showing and breeding dogs, which has continued to be a most enjoyable hobby. The sixth generation of Sarabande Shelties now lives happily with the author. Two have completed championships, several others are pointed, and many have finished obedience titles.

Her interest in dogs as a hobby, and dogs as a career in research, eventually led Phyllis into veterinary school. It seemed to be a natural extension of both interests. She graduated from Colorado State University in 1977 and has been in small animal practice since that time. Reproductive problems and breeding management have continued to be her specialty.

Phyllis participated as a lecturer at the Colorado State University short course for breeders and veterinarians on canine reproduction during the two years it was presented. Most recently, she contributed a chapter on vaginal cytology to the second edition of *Current Therapy in Theriogenology* (W.B. Saunders Co.)

*Dr. Holst pictured handling her Ch. Sarabande Bold Woodsman. "Teddy," at age 10, was chosen Best Veteran Dog at the 1984 National Specialty of the American Shetland Sheepdog Association.*

# FOREWORD

At last, a book on canine reproduction that brings to the serious breeder a factual scientific treatment of a very complex subject. It is not a definitive work, as I am sure Dr. Holst would be the first to agree, because not all of the answers are available. There is, however, a wealth of information in the literature which can be of great assistance to the serious breeder and it is fitting that the person who has brought it together for us has done some of the original research while pursuing a Masters degree, has the clinical experience in dealing with the problems in her Veterinary practice, and as a practical dog breeder understands the needs of the lay person struggling to produce that ever better litter of pups.

If this book did nothing more than dispell some of the erroneous views so prevalent in the dog world about proper time to breed the bitch, mixed litters, potency, or the administration of hormones, it would justify its existence. Fortunately for us it goes far beyond and explores the problem starting with the basic development and anatomy of the male and female reproductive system, the interaction of the various hormones culminating in sperm and egg production and their union to start new life. The complexities of the estrous cycle are covered from behavioural patterns, physical signs and by microscopic study of vaginal smears in a thorough manner including photographs taken through the microscope to facilitate our knowing what to observe. The development of the pup is reviewed from conception to weaning—explaining the normal process; discussing problems that arise, how to recognize them, what to do about them; when to call your veterinarian; and how to safeguard the bitch and her pups.

In addition, much practical information is provided regarding the mating ritual and sex act; artificial insemination, when and when not to resort to hormone therapy; nutrition during pregnancy and lactation and puppy supplementation.

The final chapters deal with the medical problems and treatment of the young pup and some of the difficulties encountered in both the male and female reproductive systems of adults requiring treatment or rendering them worthless as breeding partners.

Dr. Holst has wisely not avoided the use of scientific and technical language. To have done so would have perhaps made easier reading but at the too high price of accuracy. She has, however, come to our rescue with a comprehensive glossary of terms facilitating our understanding of this complex subject.

This is a book that should be in every breeder's library.

Quentin N. LaHam, Ph.D.
Ft. Myers Beach, Florida
May, 1984

Chapter One

# SO, YOU WANT TO BE A BREEDER!

## Introduction

For someone like myself who was never very good at drawing pictures, painting, or making figures out of clay, who never seemed to be able to catch on to music and could only sing a tune after hearing someone else sing it first—where is that creative outlet that we all need? I found it to a certain extent as a youngster when I learned to knit. After a few years of practice I could take virtually any pattern, follow the written directions, and turn a simple strand of yarn into something truly lovely, warm, colorful, and useful. But there was some limitation to that. I was only following the pattern that had already been written, recreating a sweater or a mitten that had already been created. I still love to knit; it is orderly and beautiful, and when a new project is finally finished, I feel very good. But, after I have done the most complex pattern, then what? One answer for me and many others is the ''art'' of breeding dogs. Here is a creative project with unlimited possibilities for originality, self-expression, constant challenge, new learning, new experiences, and, as a bonus, new friends.

However, there is a very important difference between creating with dogs and creating with pen, brush, yarn, or clay. The difference is this: We are not beginning the project with a brand new, never-touched ball of yarn. The dogs that are the subject of our craft are already made—they come from somewhere. Our task is not to create a dog from an amorphous mass of flesh and hair, but to make a better dog from the dog we have. And the dog we have is the end product of our own, or more likely, someone else's efforts in the past to create a better dog from the dog at hand. Just think what it means to have a purebred dog. It is a purebred dog for only one reason: because some- one cared and supervised the birth of every single generation. For decades past, or even for hundreds of years in some breeds, someone saw that the right dog was bred to the right bitch, and reared the pups to a healthy adulthood. Think how hard it is to keep a bitch in heat from cavorting about with any cur that passes by. Do you think that she cares about the lineage of her whelps? No! Humans care! And ever since the breed was originated, humans have supervised every step to improve on the features that make every breed useful, special, and unique.

So, we are involved in the art of breeding dogs. What criteria should we use in making the decision to breed or not to breed? First, we must be aware of our responsibility to keep our breed pure and strong and correct. We must be aware of the love and devotion that are behind the dogs with which we are working today. Second, we have a responsibility to learn about and to study our breed and to know what is correct. It is no excuse to say "I haven't been in the breed long enough yet to know what a good shoulder is." You must learn what a good shoulder is, how to recognize it, and where to find it. Until you know, you have no right to tamper with the breed in which you are interested. There is no excuse for ignorance. Many learning resources are available, including breed and all-breed clubs, maga- zines, and hundreds of books, as well as personal contact with experi- enced people.

**Fig. 1.1** *Dogs selected for breeding must have excellent temperament, no exceptions, and the special qualities for which the breed is prized.... (Photo by Bobbie Christensen, Rocky Mountain Training Kennel.)*

A third requirement is that a dog be of excellent quality. And, once again, no excuses for ignorance. You must know what you have. You must be able to objectively evaluate your dog and know his or her strengths and weaknesses in relation to the breed's standard of perfection. Of course, no dog is perfect, but some are close to the ideals toward which we strive. Others are far removed. There is no need to bother with a dog that is not an excellent example of the breed. Such a dog may have a place in this world, if he is cute, pretty, fun, friendly, or talented, but if he is a poor example of his breed, the genes should not be propagated.

Fourth, regardless of any specific breed characteristics, a dog considered for breeding must be in excellent health and be free of hereditary defects. He must have an excellent temperament—no exceptions. There can be no ifs, ands, or buts in the areas of health and temperament. Do not be fooled into thinking that *this dog* is so outstanding in head, or coat, or whatever, that he can be forgiven his spooky or aggressive nature. Your breed does not need that. Just look at the AKC statistics on registration for your breed. Plenty of dogs are being whelped and registered, and you can afford to be very fussy with breeding stock. Remember—as much as you may love your dog, he came from somewhere, and chances are there are many others around with the good features you prize which *also* have the temperament

**Fig. 1.2** ...*It is the only way to produce those special puppies we long for. (Photo by Bobbie Christensen.)*

correct for the breed. He is not an original creation made by you out of nothing.

A fifth consideration in deciding whether to breed is your own commitment to the puppies. Raising dogs is a time-consuming, demanding, and expensive project. But, like everything else, the rewards are there at times in equal measure. Be sure that this is for you and that you have the needed time to devote to it. If you can answer yes, then by all means go ahead, and enjoy!

This book is written for the many people I have known through the years who *do* want to breed dogs, who *are* serious about it, and who *do feel* the kind of commitment to their breed which will enable them to make a contribution. I have said that there is no excuse for ignorance about your breed and its standard and what constitutes a good dog. Unhappily, in the past, there has been an excuse for ignorance about the biology of reproduction and the technical aspects of breeding dogs. Information was not readily available. This book should help to eliminate that area of ignorance.

Ignorance leads to misinformation, erroneous assumptions, and classic "old wives' tales" about reproduction. Just as philosophers concluded that the world was flat before Columbus proved it to be round, early biologists believed that each sperm cell contained a complete miniature creature of its species curled up in its head, ready to begin growing after contact with the ovum. That idea was also proved incorrect. Yet in both instances the truth was not immediately accepted by all. Many years passed before the shape of the Earth was commonly accepted.

The information in this book has been derived from the most recent scientific research available. Some of it contradicts commonly believed notions about dog breeding, such as the possibility of pups of different ages in a single litter. Read it carefully, and it will become clear that the reproductive processes in dogs are just as orderly, just as intricately controlled, as the rotation of a round planet Earth around the sun. There is, of course, much that we still don't understand, but the curiosity of the many talented scientists currently working in research on canine reproductive biology will constantly expand our knowledge. Ultimately all the gaps in our knowledge will be filled.

Chapter Two

# WHERE PUPPIES COME FROM

## Development and Anatomy of the Bitch

The basis of our understanding of reproductive processes in dogs is first to know the parts of the reproductive system. Knowing the parts that make up the whole is fundamental to really understanding anything around us. Sure, you can drive your car even if you don't know exactly what every part does in the engine, but you'd better know a thing or two in order to properly care for the car and avoid breakdowns. Likewise you can oversee the breeding of a litter of pups if you don't know every part of the bitch's reproductive system, but if you do know and understand those parts, your success in dog breeding will be greatly enhanced. Figure 1 will help introduce some terminology that will be used throughout the text.

The bitch's reproductive organs are the same as in any other mammal, including humans. The actual design, however, is specially modified to enable her to produce litters of several offspring at one time. The reproductive organs of the bitch include the ovaries, oviducts, uterus, vagina, and vulva.

The ovaries are the female gonads (organs that produce reproductive cells and hormones). In normal bitches there are two ovaries and they lie in a position just caudal ("toward the tail," behind) to each of the kidneys in the abdomen, close to the body wall. The two ovaries develop in the embryo from primitive, undifferentiated gonadal tissue at just the location where they are found later in life. The left ovary is always located just a little bit farther caudally than the right, which is also true of the left kidney compared to the right. Each ovary is held in position by ligaments that extend to the body wall at about the level of the last rib. The ovaries' primary purpose is to produce cells—the ova, or eggs—which after being released (ovulated) and then fertilized by the male's sex cells—the sperm—produce new members of the species. Sounds simple enough, but much is involved to ensure that the goal is achieved.

Each ovary is a small, solid, rounded structure made up of two layers—an inner area called the medulla and an outer rim called the cortex. The medulla is made up of supportive tissue, blood vessels, nerves, and lymph channels. The cortex is the main functional part of the ovary and is made up of connective tissue containing germ cells (primitive ova), follicle cells (cells that surround, support, and nourish the ova), and follicles (ova surrounded by fluid and follicular cells). In embryonic life the ovary is covered by a solid sheet of germ cells. Rows of germ cells migrate deep into the ovary and are organized into primitive follicles. Huge numbers of follicles begin to develop that later degenerate. It has been estimated that the newborn bitch puppy has 700,000 ova in her ovaries. Most of these degenerate before she even reaches puberty, and further losses continue throughout her life.

Ovulation is the process of releasing a mature ovum from a mature follicle so that it can be joined with a fertilizing sperm cell. Ovulation occurs only when the bitch is in season, which will be discussed in considerable detail later. A primordial follicle that is destined to produce a normal ovum and to ovulate is at first a solid spherical mass of follicular cells surrounding the ovum. The follicle enlarges, first

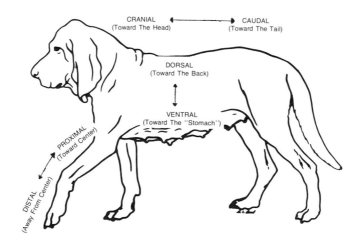

**Fig. 2.1 Relative Orientation of Anatomical Structures.** *These terms will be used repeatedly throughout the text.*

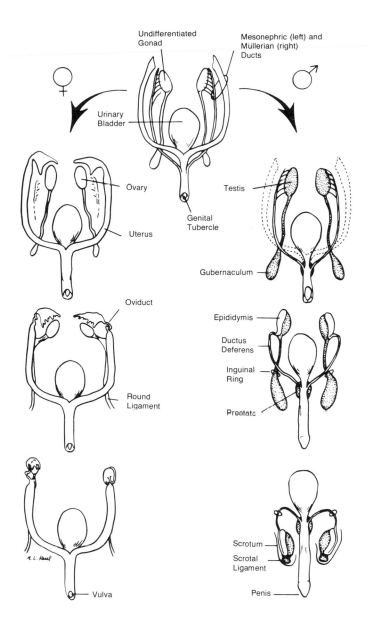

**Fig. 2.2 Development of Female and Male Reproductive Organs from Primitive Undifferentiated Sex Organs and Duct Systems.** *In the early embryo two duct systems and undifferentiated gonads are found. The ovaries and Mullerian duct system develop in the female, while testes and the mesonephric duct system develop in the male. The ovaries remain for life at the site of their development, but the testes migrate to a final location in the scrotum.*

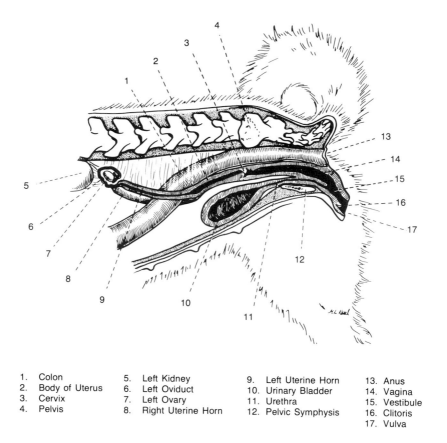

| | | | |
|---|---|---|---|
| 1. Colon | 5. Left Kidney | 9. Left Uterine Horn | 13. Anus |
| 2. Body of Uterus | 6. Left Oviduct | 10. Urinary Bladder | 14. Vagina |
| 3. Cervix | 7. Left Ovary | 11. Urethra | 15. Vestibule |
| 4. Pelvis | 8. Right Uterine Horn | 12. Pelvic Symphysis | 16. Clitoris |
| | | | 17. Vulva |

**Fig. 2.3 Reproductive Organs of the Bitch.** *The right and left ovaries lie near the caudal end of the corresponding kidneys. The uterine body, cervix and vagina lie ventral to the colon and dorsal to the urinary bladder.*

by proliferation of follicle cells and later by development of a fluid cavity in the follicle. After the fluid-filled cavity develops, the follicle is referred to as a Graafian follicle. The ovum lies along one side of the Graafian follicle, surrounded by a special cluster of follicle cells. As the follicle enlarges, it approaches the surface of the ovary. Its outer wall becomes thinner in preparation for ovulation. Eventually, when the follicle is fully mature and after some important hormonal events, the follicle ruptures and the ovum is released. Rupture of the follicle and release of the ovum constitutes ovulation.

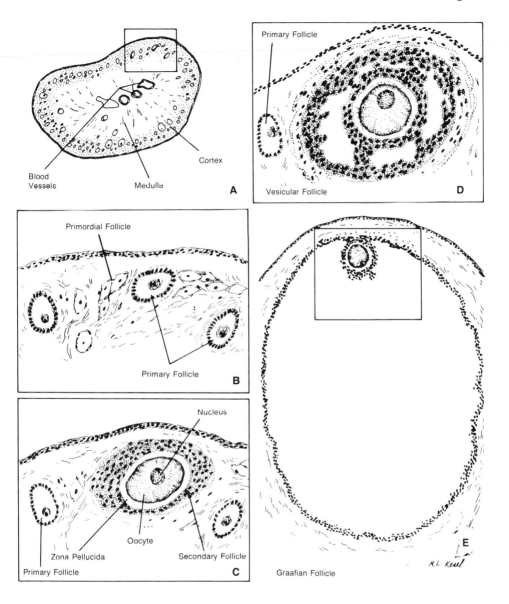

**Fig. 2.4 Development of Follicles in the Ovary. A, B.** *Cross-section of a pre-pubertal ovary. Primordial follicles consist of an oocyte surrounded by a single layer of flattened follicular cells. A primary follicle contains a single layer of cuboidal follicular cells.* **C.** *A secondary follicle is the oocyte with several layers of follicular cells.* **D.** *Vesicular follicles develop under the hormonal stimulation prior to each estrous cycle. Fluid filled spaces develop as the follicle enlarges.* **E.** *A Graafian follicle consists of a large fluid cavity lined with follicular cells.*

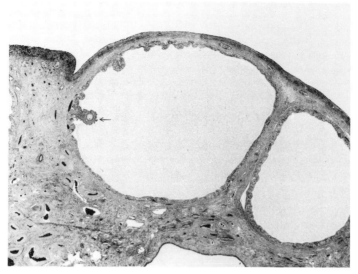

**Fig. 2.5 Photograph of a Thin Section of Ovary Approximately Two Days Prior to Ovulation.** *Two Graafian follicles are shown, each lined with a thin lining of follicular cells. The oocyte (arrow) lies along one side surrounded by follicular cells, and still attached to the follicular wall.*

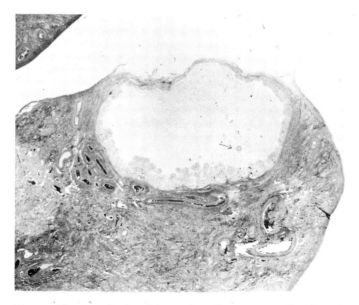

**Fig. 2.6 Photograph of a Thin Section of Ovary Immediately Before Ovulation.** *The follicular lining cells have begun the process of luteinization and are already producing progesterone. The oocyte (arrow) is nearly freed from its attachments, and the outer wall of the follicle is thinned in preparation for ovulation.*

Following ovulation, the ovaries take over another extremely important function. The follicular lining cells proliferate and change both in their structure and function. Each empty follicle becomes a new gland, the corpus luteum (CL), which means "yellow body." Each CL is a solid mass of cells that produce mainly progesterone, the hormone that maintains pregnancy.

The oviducts are small, relatively long, thin tubes whose purpose is to carry the ova from the ovaries to the uterus. During their transit through the oviducts, the ova are prepared for fertilization, are fertilized, and undergo the first stages of early embryonic development. There are two oviducts, one adjacent each ovary. Each lies within a pouch of thin membranous connective tissue, the ovarian bursa ("purse"), which completely surrounds the ovary. So, when the ova are ovulated, there is not really much chance that they will be lost by being released free into the abdomen. The end of the oviduct is made up of numerous frondlike folds of tissue (the fimbriae) which actually lie right on top of the ovary. The ova are swept by movements of fluid into the oviduct. At the time of ovulation, the fimbriae are large and the fluid in the bursa abundant, helping to ensure capture of the ova by the oviduct. I have seen, at the time of spay, a few cases in which the bursa was not properly formed, and the ovary lay fully exposed in the abdominal cavity. Presumably any ova produced from such an exposed ovary would have been lost into the abdomen.

The cells lining the inside of the oviducts are covered with hairlike projections (cilia) that sweep rhythmically and move fluid in one direction. The same type of cilia-covered cells help keep our lungs clear of debris by sweeping it outward. Rhythmic muscular contractions and movement in the cilia move the ova quickly through the oviduct. By two days after ovulation, they are at the uterine end of the tubes, where they are fertilized. The ova stay in that location for several more days before they are released into the uterus. The ova are held in the oviducts because of a functional closure at the uterine end. High levels of estrogen, one of the reproductive hormones, causes this "tubal lock" until about seven days after ovulation.

The uterus ("womb") houses developing embryos (later to become fetuses) during pregnancy. It is a hollow, muscular organ in all mammals. In the bitch it is tubular and roughly Y-shaped. It consists of two long horns (the upper prongs of the Y), a short body connected to each horn (the stem of the Y), and a short neck, or cervix. The upper (cranial) end of each uterine horn is connected to the caudal end of its corresponding oviduct near the ovary. The length of the horns is much greater than the diameter, and the size of the uterus varies tremendously with the bitch's age, stage of estrous cycle, and pregnancy. The two horns extend to and converge at a point far caudal in the abdomen, at about the rim of the pelvis, there forming the body of the uterus. The body has the same diameter and structure as each of the horns.

The wall of the uterus is made of three layers: an outer, thin covering of peritoneum called serosa; a muscular portion made up of two layers—an outer longitudinal layer and an inner thicker circular layer; and the thickest layer of all, the endometrium, or inner lining. The endometrium contains glands and a rich supply of blood vessels. The endometrium undergoes cyclic changes depending upon the stage of the bitch's estrous cycle, the most dramatic of which involves implantation of the developing embryos during pregnancy.

The cervix is a firm, fibrous/muscular structure that serves as the channel from the uterus into the vagina. It consists of a thickened protrusion that is tipped at an angle directed caudally and downward (bitch standing) from the uterus into the vagina. It is normally hidden from view from the vaginal side by a fold of vagina. Because of its peculiar position and anatomy, the cervix cannot usually be seen

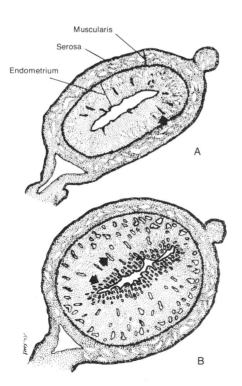

Fig. 2.7 **Structure of the Uterus. A.** *The uterus during anestrus consists of a relatively thin endometrium with poorly-developed glands (arrow).* **B.** *During estrus the endometrium is thickened and development of deep and superficial glands is dramatic (arrows).*

through the vagina. It can rarely be entered, as for insemination, culture, or treatment, except during and shortly after whelping, when it is dilated. It is open during estrus, of course, which allows for the entrance of sperm, but the opening is so small that most instruments and culture swabs cannot be inserted.

The vagina in a bitch is quite a large, expandable muscular/membranous canal ("birth canal") extending from the cervix to the vulva. The cervix may protrude a short distance into the vagina, as mentioned previously. Longitudinal (lengthwise) folds are present throughout the length of the vagina, along with some transverse (crosswise) folds, and these allow for tremendous expansion during mating, pregnancy,and whelping. The vagina is lined with a stratified (layered) squamous (flattened) epithelium (lining of cells) which changes in thickness and nature during the bitch's cycle. The vagina opens caudally into the vulvar area just ahead of the urethral opening (the outflow tract from the urinary bladder). It is demarcated from the vulva by a ridge, but a hymen is not normally seen in the bitch. From time to time a fibrous stricture or a band of fibrous tissue across the opening may be encountered, and that could be significant at breeding time.

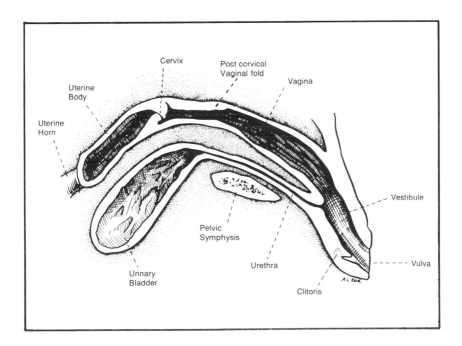

**Fig. 2.8 Detail of the Cervix, Vagina, Vestibule and Vulva.**

The vulva is the external genital organ of the bitch, and it consists of three parts: the vestibule, the labia, and the clitoris. The vestibule is the channel or space connecting the vagina to the outside. On the ventral floor of the vestibule, just caudal to the vagina, is the opening of the urethra, from the bladder. It is demarcated as a small elongated bump or tubercle.

The labia, or lips of the vulva, form the external boundary of the vulva. The labia are soft and pliable and join ventrally to form a pointed projection that extends downward and backward from the body. The size of the labia varies tremendously, depending upon the stage of the bitch's estrous cycle or pregnancy.

The clitoris is analogous to the male's penis and is a small, pointed structure that lies along the ventral floor of the vestibule. It is made up of fat, elastic connective tissue, and a small amount of erectile tissue.

## Mammary Glands

The mammary glands are actually modified skin glands. The bitch (and dog actually) has four to six pairs, usually five pairs, located in two rows along the length of the ventral abdomen. It is not unusual to have some asymmetry in the placement of the teats, or nipples, or to have one or more teats missing. The glandular tissue develops during pregnancy and is extremely active during lactation, then regresses to an inconspicuous amount after weaning. The glands empty to the outside through the teats in a cluster of ducts, much like a sieve. There may be eight to twenty-two ducts at the end of each teat. This pattern is quite different than that seen in the cow, where each teat has a single large duct opening.

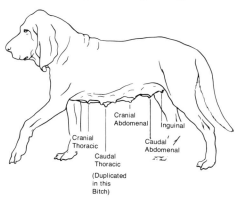

**Fig. 2.9 Arrangement of the Mammary Glands of the Bitch.** *Normally five symmetrical pairs of glands are present, but four or six pairs are normal. Missing or duplicated glands are also commonly seen.*

Chapter Three

# MORE WHERE PUPPIES COME FROM

## Development and Anatomy of the Male

The reproductive organs of the male can be divided into four components according to their role in the production and delivery of semen. Semen is the fluid that contains glandular secretions and the male sex cells—the spermatozoa, or simply sperm. It must be deposited in the female's reproductive tract at just the right time for reproduction to occur.

The gonad in the male is the testis. The two testes are the site of production of the male sex hormones and of spermatozoa.

A duct system transports the spermatozoa from the testis. The spermatozoa mature in the ducts and are mixed with fluid before leaving the male's body. The epididymis, vas deferens, and urethra make up the duct system.

The dog has one secondary or accessory sex gland—the prostate. The prostate produces fluid that is mixed with the spermatozoa, testicular, and epididymal fluids and constitutes a portion of the dog's semen.

The external sex organ, the penis, with its cover, the prepuce, is responsible for physically delivering the semen to the vagina during mating.

## The Testes and Sperm

The testes, the manufacturing plant for sperm and hormones, are made up of tubules and connective tissue that supports and organizes

**16**

the tubules into a coherent organ. Among and within and tubules are secretory cells that produce various reproductive hormones.

Each testis is surrounded by a capsule of fibrous connective tissue, and partitions of connective tissue divide the organ into lobules. Each lobule contains long but tightly coiled seminiferous tubules, which give the organ a spongy appearance when cut. Spermatogenesis, the production of spermatozoa, takes place inside the seminiferous tubules. In a normal dog, sperm production occurs constantly, rather than seasonally, as is the case in some other animals. The seminiferous tubules of each testis connect at their distal ends with straight tubules—or one might say that the tubules become straight at their distal ends. The numerous straight tubules then form a honeycomb of passages that empty into small collecting ductules—the efferent ductules. The collecting ductules become the head of the epididymis, a tortuous network of tubules in which sperm maturation occurs. Finally each epididymis empties into, or continues as, the ductus deferens.

The cells lining the seminiferous tubules are collectively referred to as the germinal epithelium. Two types of cells are present: the germinal cells, which actually give rise to new spermatozoa, and Sertoli cells, which support the germinal cells and produce estrogens

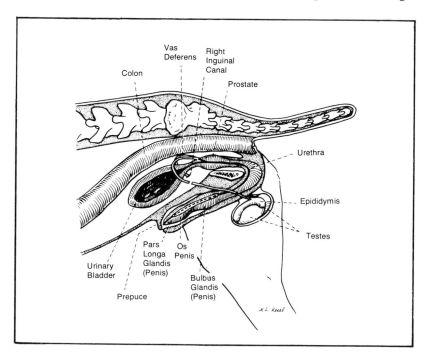

**Fig. 3.1 Reproductive Organs of the Dog.**

and other hormones in small amounts. Lying between the tubules, usually in clusters, are the interstitial or Leydig cells. Their primary function is to produce testosterone, the androgen which is important to the proper functioning of the germinal epithelium, to secondary sexual characteristics, and to sexual performance. Leydig cells also produce other androgens, progesterone and probably estradiol.

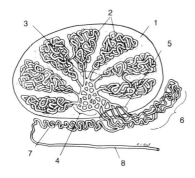

1. Capsule (Tunica Albuginea)
2. Seminiferous Tubules
3. Straight Tubules
4. Rete Testis
5. Efferent Ductules
6. Head of the Epididymis
7. Tail of the Epididymis
8. Ductus Deferens

**Fig. 3.2 Structure of the Testes.** *Spermatozoa are manufactured within coiled seminiferous tubules and transported and matured in the duct system.*

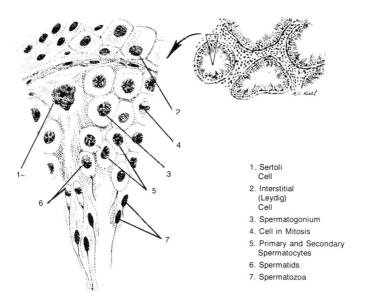

1. Sertoli Cell
2. Interstitial (Leydig) Cell
3. Spermatogonium
4. Cell in Mitosis
5. Primary and Secondary Spermatocytes
6. Spermatids
7. Spermatozoa

**Fig. 3.3 The Seminiferous Tubule.** *Spermatozoa are produced in the tubules from undifferentiated germinal cells (spermatogonia). The interstitial (Leydig) cells produce testosterone, and Sertoli cells produce estrogen and androgen-binding protein.*

Inside the seminiferous tubules, sperm are produced from undifferentiated germinal cells called spermatogonia. The spermatogonia lie in the deepest layer of the lining of the tubules. They divide by mitosis to produce a population called terminal B spermatogonia. Spermatogonia each contain a full double (diploid) set of chromosomes. The chromosomes are duplicated, and then two meiotic divisions occur, producing four spermatids. Each of the four spermatids has a single (haploid) set of chromosomes and goes on to become a mature spermatozoon by a kind of metamorphosis. At first the spermatids are poorly separated, and they lie deep within recesses of the elongated Sertoli cells. They gradually move toward the tubular lumen while evolving into spermatozoa with fully differentiated heads, midpieces, and tails. The spermatozoa line the seminiferous tubules, heads pointed outward toward the periphery, tails inward toward the lumen, until their release into the lumen of the tubule.

At any one time, sperm are in various stages of maturation at different places along the seminiferous tubules. Thus, mature spermatozoa reach the lumen continuously.

The time required for a full cycle of spermatogenesis, that is, for an undifferentiated primitive germ cell to divide and differentiate into fully developed spermatozoa, is roughly six weeks in several other domestic animals, nine weeks for bulls. The time required for spermatogenesis in dogs is reported to be sixty-two days. Adding an estimated ten to fourteen days migration time through the epididymis, you can understand that in examining a given semen sample you are seeing sperm that have been developing and maturing for at least ten weeks. Any damage to the sperm caused by infection, abnormal temperatures, chemical insults, or irradiation may not be detected in the semen until ten weeks after the event. Likewise, abnormalities seen today may reflect insults to the process of spermatogenesis that occurred weeks before. In fact, three to five months may be required for recovery from a problem that interferes with spermatogenesis.

Spermatozoa and seminal fluids are passed through the tail of the epididymis, through the vas deferens and the urethra, and into the female's vagina during ejaculation. Spermatozoa have been shown to reach the female's oviduct, the site of fertilization, within twenty-five seconds of ejaculation. Considering the distance traveled and the distance that a spermatozoon is actually able to swim in a given period of time, the swimming of spermatozoa could not possibly carry them to the oviduct so quickly. The peristaltic contractions of the uterus are what carry the sperm to the oviduct in such a short time.

Sperm capacitation time—the time after ejaculation required for a spermatozoon to be matured or changed chemically so that it is able to fertilize an ovum—has been reported to be seven hours in the dog.

Secretions from the female tract, possibly the follicular cells, may be involved in the chemical process of capacitation. After breeding, the spermatozoa can be found in large numbers lying with their heads directed inward deep within uterine glands. It appears that the spermatozoa may be nourished by secretions ("uterine milk") from these glands. Whether these particular spermatozoa move later into the oviduct or can have a part in fertilization is unknown. In microscopic sections, very few spermatozoa are seen in the oviduct, compared with many in the uterus.

Studies have been conducted in which a single mating was allowed and the time of ovulation and subsequently fertilization determined. The lifespan of spermatozoa within the female's reproductive tract was shown to be at least seven days. In other words, fertilization regularly occurred from a single mating seven days before mature ova were ready to be fertilized in the oviducts. Other work has shown that motile sperm can be found in the female tract ten to eleven days following mating. Whether they would be fertile after that length of time was not determined. Knowing these special characteristics of canine spermatozoa is essential to our understanding of the reproductive processes in dogs.

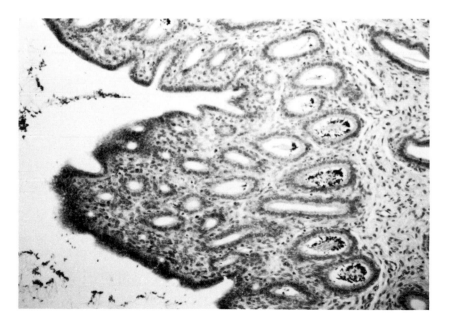

**Fig. 3.4 Thin Section of the Endometrium Following Breeding.** *Large numbers of spermatozoa are found free in the lumen and deep within the endometrial glands. It is believed that the spermatozoa are nourished by secretions from the glands.*

# Descent of the Testes

The testes form in a location just caudal to the kidneys, just as do the ovaries. Then during the last days of gestation and first days after birth they migrate caudally through the inguinal rings, out of the abdomen, and into the scrotum. At about eight weeks postbreeding, the testes of the fetus are located in the abdomen about halfway from the kidneys toward the inguinal canal. At birth they should be near the inguinal ring or in or just beyond it. By two weeks of age the testes should be fully descended into the scrotum. That is not to say that they are easy to feel in a two-week-old male. Sometimes it is difficult because they are very small. But they ought to be there and ought to stay put. In a normal male, it should be possible for an experienced person to find the testes in the scrotum at six weeks of age.

The descent of the testes from the abdomen into the scrotum takes place because of differential growth of ligaments and supporting structures, mainly the gubernaculum. This structure connects the caudal pole of the testes to the inguinal region. Late in fetal development, the gubernaculum grows in length and diameter and expands beyond the inguinal canal into the scrotal sac. Then, because of traction, the intra-abdominal part of the gubernaculum is drawn into the extra-abdominal part, and in the process, traction is exerted on the testes. The testes are then drawn into the inguinal canals and down into the scrotum. The gubernaculum decreases tremendously in size and becomes the proper ligament of the testis and the ligament of the tail of the epididymis, which hold the respective structures in the scrotum. The fetus is relatively small during the descent of the testes, so the distance of the actual migration is not far and the structures are very small at the time.

All other structures of the male reproductive system, including epididymis, vas deferens, urethra, prostate gland, penis, and prepuce, are fully developed, although they are small and immature, at the time of birth.

# Cryptorchidism

It would be extremely rare for a male to be born with only one testicle, a condition that is implied in the term "monorchid" (one testis). The gonads are such a basic and important part of an animal that their development seems to be less subject to error than some other organs. The term, therefore, should not be used unless and until it has been determined by exploratory surgery that the individual in question does indeed only possess one testis. What most people mean when referring to a male as "monorchid" is that he has only one testicle located where it belongs in the scrotum. "Cryptorchid" (from Greek *kryptos*, meaning hidden) is the term properly used to describe the male with the missing testicle. In virtually all cases, the missing

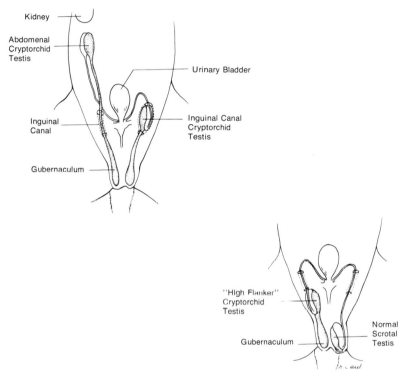

**Fig. 3.5 Cryptorchidism** *is an arrest in the normal migration of the testes from their site or origin near the kidneys to the scrotum. The cryptorchid testis may be found in the abdomen, in the inguinal canal, or under the skin in the flank region.*

testicle is either located under the skin of the flank between the scrotum and the external inguinal ring (a "high flanker"), or in the inguinal canal, or in the abdomen. When one testicle is in an abnormal location, the individual should be referred to as a "unilateral cryptorchid." He is a "bilateral cryptorchid" when both testicles are in an abnormal location.

The cryptorchid testis, because of a higher than normal temperature, is not fertile, although it will generally produce the male hormones. The cryptorchid testis, regardless of location, is always smaller than the normal testis. The temperature in the scrotum is lower than the rest of the body, and the lower temperature is essential for sperm production. The unilateral cryptorchid still ought to be fertile because he has one normal testis. However, he should not be bred, even if he is fertile, because the trait tends to be heritable. The American Kennel Club does not allow cryptorchid dogs to compete in breed competition. Furthermore, the cryptorchid testis is prone to develop cancer later in the dog's life. A cryptorchid dog should not be considered a sound individual and ought to be neutered.

Various theories have been proposed to explain the mode of inheritance of cryptorchidism in the dog. Probably the most commonly held theory is that the trait is recessive, because it appears in litters from normal parents. It has been suggested that the trait is sex-linked, because it can obviously only be expressed in males. If this is true, a female may be considered a carrier and be capable of producing either carrier daughters or affected sons. Others have even suggested that it is dominant trait, perhaps with variable expression. Considering the nature of the descent of the testes, with the several structures and several processes and changes involved, the trait may well be polygenetic.

Slow descent of the testes may be seen in some breeds or individuals. This probably has a genetic basis but in some cases may be nothing more than normal, small testes that are difficult to palpate. Testes that seem to appear and disappear are usually small in size and probably have loose ligamentation that allows them to slide back and forth from the scrotum to the inguinal region.

# The Duct System

Since abnormalities of the epididymis, vas deferens, and urethra are so rare, I will not give a lengthy discussion of their development. Every embryo has a primitive kidney, consisting of a system of tubules and ducts, which is gradually converted into the tubules and ducts that make up the outflow channels for the gonads. The seminiferous tubules that make up the testes connect with a series of small tubules, the efferent ducts, which empty into the epididymis.

The epididymis is a long, coiled tube that lies surrounded by connective tissue on the surface of the testis. It can be easily felt in a normal dog as a ridge or swelling along the dorsal and lateral sides of each testis. It is in the epididymis that spermatozoa are stored and matured in preparation for ejaculation. On the average, an individual spermatozoon will spend ten to fourteen days in the epididymis. The length of the epididymis and the slowness of the movement along its length allow for maturation of the spermatozoa. The cells lining the epididymis produce a secretion that is believed to nourish the spermatozoa and help to prepare them for their role in fertilization. It is also one of the constituents of semen.

The vas deferens, or ductus deferens, is a continuation of the duct of the epididymis. It leaves the tail of the epididymis and goes up toward the inguinal area in the spermatic cord. The spermatic cord is made up of the ductus deferens, the veins and arteries serving the testis and epididymis, nerves, and a covering of connective tissue. A thin band of muscle—the cremaster muscle—extends down from the abdominal wall into the scrotum parallel to the spermatic cord and attaches to the tunics of connective tissue near the testis. Contraction and relaxation of this muscle are mainly responsible for the carriage of the testes either high, close to the body, or lower in a more pendulous position, depending upon the ambient temperature at the time.

The vas deferens enters the abdominal cavity through the inguinal canal, lies along the body wall surrounded by connective tissue, curves toward the midline, and enters the prostate at the neck of the urinary bladder.

The two deferent ducts pass through the prostate and empty through two small slits into the urethra at a point inside the prostate. Beyond this point, the dog's semen and urine share a common exit passage through the urethra.

# The Prostate

The prostate is the dog's only accessory sex organ. It is a more or less symmetrical gland approximately the size of a walnut in a medium-sized dog and consists of two lobes separated by a shallow cleft along its dorsal surface. It completely surrounds the urethra and is located just caudal to the neck of the bladder. If the bladder is full of urine, the bladder and prostate may lie forward of the pelvis in the abdomen. When the bladder is empty, the prostate will normally lie along the floor of the pelvic canal. In this location, it can be felt in digital (finger) examination, and its size, shape, and texture can be evaluated.

The prostate is made up of glandular tissue enclosed in connective and smooth muscle tissue. It has a thick, fibrous/muscular capsule, and the connective tissue extends into the organ, dividing it into lobules. The prostate produces a thin, colorless, watery secretion that is ejaculated at the time of breeding, mainly after the sperm-rich fraction has been ejaculated. The exact function and importance of the prostatic fluid in fertilization is not clearly understood. It probably thins and increases the volume of the ejaculated semen, but beyond that, its importance is uncertain.

The urethra is the tubular outflow channel from the bladder to and through the penis. It has no particular structural peculiarities of note but is lined with muscle fibers which by their rhythmic contractions help move the contents to the outside.

# The Penis and Prepuce

The penis is the male copulatory (breeding) organ, and it is made up of three principal parts: the root, the body, and the distal, free portion (glans). The glans is divided into the bulb (bulbus glandis) and the pars longa glandis, which doesn't really have a simpler common name. The root and body of the penis are firmly attached to the ventral body wall, while the glans is free from the body wall but is entirely enclosed within the prepuce (except in erection). Deep within the glans is a bone, the os penis, which gives the structure support, especially during the early stage of copulation, or mating. This structure is found only in dogs and cats among the common domestic animals.

The main mass of the penis is made up of an elaborate network of blood sinuses surrounded by connective tissue. During erection, the tissue fills with blood and the organ enlarges tremendously. The bulbus glandis enlarges to a roughly spherical shape, a conspicuous and unique feature of anatomy in the male dog. It is the enlargement of the bulb within the caudal part of the vagina during mating that is responsible for the coital lock, or tie, characteristic of dogs. The pars longa glandis also enlarges during erection but does not change in its basic shape. The portion of the penis immediately behind the bulb has relatively much less cavernous erectile tissue. During breeding, it is at this area that, when the male dismounts, the penis is able to twist 180 degrees without discomfort to the dog or damage to the tissues. The penis in this area is relatively fibrous and elastic.

The urethra runs as a single straight tube through the entire length of the penis, at a level below most of the erectile tissue and beneath the os penis. It exits as a simple round hole at the ventral edge of the tip of the penis.

The surface of the penis is covered with a thin, delicate skin that is moist because of its constant protection by the prepuce. The same type of skin lines the inner surface of the prepuce, which is the tubular sheath that contains and protects the penis. The skin lining the inner surface of the prepuce has a number of lymph nodules that may become inflamed and enlarged when there is infection in the sheath.

## Mechanism of Erection

When the nervous stimulation (sexual excitement) for erection occurs, blood flow into the penis increases, and blood in large quantities expands the spaces of the sinuses and cavernous erectile tissue. Blood pressure in the area increases, and there is a partial inhibition of drainage through the veins leaving the area. Therefore, more blood enters the organ than leaves, and the erectile tissue enlarges. It is uncertain just what factors determine the length of time the erection is maintained. The length of the coital lock is extremely variable. In years past, it was believed that a muscular sphincter in the bitch's vagina contracted and gripped the penis behind the bulb and that the tie would be maintained as long as the sphincter was in contraction. In more recent years, however, it has been realized that it is simply the size of the bulb of the penis in the caudal portion of the vagina that holds the penis in place, not a muscular sphincter. Perhaps (and this is conjecture), the relative size of the male and female determines the length of the tie. I have observed in one of my own stud dogs that the length of the tie varies from five to eighty minutes, but that with a given individual bitch it has been consistent. Some bitches tie for fifteen to twenty minutes in each of two to three breedings, others thirty to thirty-five minutes, and a few sixty minutes or more. It is unlikely that any harm would result from even the longest ties, although in extreme cases the penis might be damaged because of compromised blood flow.

26

Chapter Four

# HORMONAL CONTROL OF REPRODUCTION

The reproductive functions of both the male and female are controlled by hormones, the brain being the site of the ultimate control. A portion of the brain, called the hypothalamus, is where a certain chemical substance referred to as gonadotropin-releasing hormone (GnRH) is produced. GnRH is released from cells of the hypothalamus, travels to the pituitary gland (located adjacent to the hypothalamus at the base of the brain), and stimulates the cells of the pituitary gland to release their products. The pituitary gland is a complex organ that produces several hormones, including those for growth, thyroid stimulation, adrenal cortical stimulation, and reproduction. The reproductive hormones include follicle-stimulating hormone (FSH) and luteinizing hormone (LH) (also called interstitial-cell-stimulating hormone (ICSH) in males). Both males and females have FSH and LH, and the hormones are the same in each sex, even though their names tend to imply functions that are female in nature.

## Male Hormonal Functions

In the male, GnRH stimulates the pituitary gland to release FSH and LH, which in turn act specifically on the testis to stimulate the production of spermatozoa and hormones. There are three target sites for FSH and LH in the testis:

1. The seminiferous tubules, which are responsible for producing spermatozoa from immature germ cells.
2. Sertoli cells, large cells that lie within the seminiferous tubules among the germ cells. Sertoli cells secrete estradiol and androgen-binding protein.

3. Interstitial cells, or Leydig cells, which secrete testosterone, the major male sex hormone, and which are located in the connective tissue around seminiferous tubules.

FSH is responsible for stimulating the seminiferous tubules and the Sertoli cells. Under the influence of FSH, spermatozoa are produced in the tubules, and estradiol and androgen-binding protein are produced by the Sertoli cells. Estradiol is an important female reproductive hormone, and its function in the male is not well understood. The androgen-binding protein is released into the lumen of the seminiferous tubules and apparently helps to hold testosterone and to increase its concentration in that area.

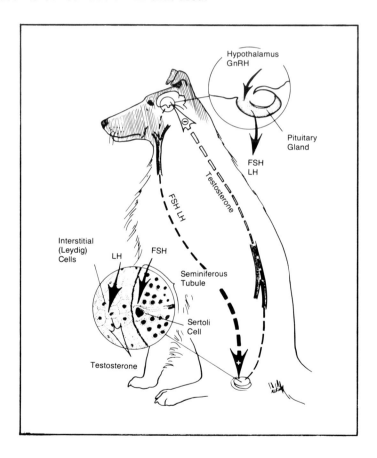

**Fig. 4.1 Regulation of Reproductive Functions by Hormones in the Male.** *GnRH stimulates the pituitary gland to produce FSH and LH. FSH and LH stimulate three target cells in the testes and are in turn regulated by a negative feedback mechanism by levels of testosterone in the blood. See text for details.*

LH stimulates secretion of testosterone by the interstitial cells of Leydig. Testosterone has several functions. It is essential for the development and maintenance of the sex characteristics of the male, it stimulates the male's sexual behavior, it maintains the reproductive tract and the prostate gland, and it helps stimulate spermatogenesis. Testosterone levels within the seminiferous tubules must be much higher than in the general circulation for normal spermatogenesis to occur, and the androgen-binding protein from the Sertoli cells assists in that process. After the spermatozoa are formed, they move into the epididymis; high concentrations of testosterone are also needed in the epididymis for its proper functioning.

# Female Hormonal Functions

Reproductive activity in the bitch is regulated by the same hormones that function in the dog. The final site of activity is the ovary instead of the testis, and the functional hormones are estradiol and progesterone rather than testosterone.

Gonadotropin-releasing hormone (GnRH) is produced in the hypothalamus and causes the pituitary gland to release FSH and LH. Follicle-stimulating hormone stimulates the growth of follicles in the ovaries and helps to prepare them for ovulation. FSH and LH together cause the follicular cells to secrete estradiol.

Estradiol has several functions in the bitch. It causes the outward signs of estrus to appear, including swelling of the vulva and the blood-tinged vaginal discharge. It begins to prepare the uterus for pregnancy, and it causes hairlike cilia to grow from the cells lining the oviducts. Movement of the cilia carries ova through the oviduct to the distal end where fertilization occurs, and later to the uterus. Once several follicles have been formed, the resulting high level of estradiol, along with another substance, folliculostatin, decreases the secretion of FSH by a negative feedback on the pituitary gland so that no more follicles are formed during that cycle. The secretion of LH causes the follicular cells to secrete some progesterone. Then the decreasing estradiol level, along with the rising progesterone level, causes a marked surge in LH (a fifty-fold increase), which triggers ovulation of the mature follicles. The LH also causes the follicles to be transformed into corpora lutea, the glandular structures that secrete the progesterone needed to maintain pregnancy. Estradiol, along with progesterone, causes the bitch to exhibit sexual behavior typical of estrus. Progesterone completes preparation of the uterus for pregnancy and maintains it in a quiet state suitable for the development of the newly conceived pups.

Other pituitary hormones are needed to support normal reproduction, even though they are not specifically sex hormones. Growth hormone (somatotropin) is concerned with the growth of all tissues, and thyroid-stimulating hormone (TSH) controls the release of thyroid hormones that regulate animals' metabolic rate. Adrenocorticotropic hormone (ACTH) stimulates production and release of steroid hormones from the cortex of the adrenal glands.

The posterior lobe of the pituitary gland secretes vasopressin and oxytocin. Vasopressin is not directly involved in reproduction, but the animal's general health depends on the inter-action of vasopressin with adrenal hormones to regulate water and electrolyte balances. Oxytocin is extremely important in causing contractions of the uterus and oviducts, needed for sperm transport at the time of breeding. It is also essential in stimulating contractions of the uterus during whelping. After whelping, oxytocin remains an essential component of lactation, stimulating milk letdown.

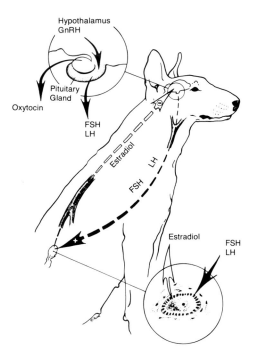

**Fig. 4.2 Regulation of Reproductive Functions by Hormones in the Female.** *GnRH stimulates the pituitary gland to produce FSH and LH. FSH and LH stimulate follicular cells of the ovaries to produce estradiol, which as several important functions during the estrous cycle. The balance of FSH and LH changes during the cycle and leads to ovulation and production of progesterone to maintain pregnancy. See text for details.*

Chapter Five

# STAGES OF THE ESTROUS CYCLE

The estrous cycle is a continuous and repeating sequence of events, each event building from the previous one and leading to the one that follows. The cycle is delicately controlled by various hormones, some originating in the hypothalamus, some in the pituitary gland, and others in the ovaries. The hormones have effects upon the cells of the reproductive organs and produce specific changes. They likewise affect cells in the brain and cause changes in behavior. The cycle is continuous, yet the stages are arbitrarily defined and described as if they were actually separate. This is done merely for our convenience in understanding these complex events. The estrous cycle is the series of events that produces ova ready to be fertilized, allows for the union of male and female gametes, and prepares and maintains the female reproductive organs to nourish and protect the new conceptuses until they are mature enough for birth. The stages of the estrous cycle are anestrus, proestrus, estrus, and diestrus.

## Anestrus

**Anestrus is a period of quiescence, or rest, lasting from two to ten months, during which the reproductive organs are inactive.** This resting period is characteristic of wild and domestic dogs, and its length varies with individuals. The type of reproductive pattern seen in dogs and characterized by regular periods of anestrus is referred to as "monestrous." By contrast, animals that cycle repeatedly without a period of rest, such as cattle, are "polyestrous." On the average, bitches have a five-and-one-half-month anestrus, which results in an eight-month cycle, or three cycles every two years. Cycles from five

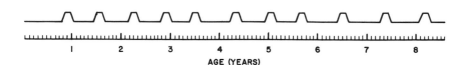

AGE (YEARS)

**Fig. 5.1 Estrous Cycles** *in a Shetland Sheepdog bitch during the first eight years of her life. The baseline represents anestrus. Each rise represents an estrous cycle and includes proestrus, estrus and diestrus, and covers a period of approximately 2½ months. A tendency to show longer anestrus with advancing age is shown. Also note that anestrus is not absolutely regular in length.*

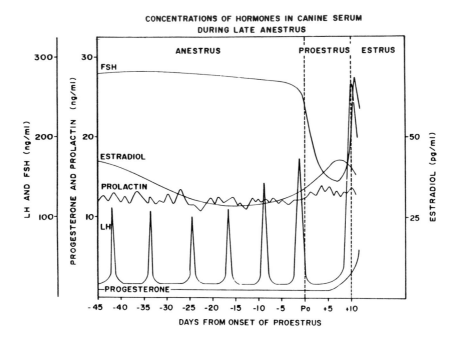

**Fig. 5.2 Levels of the Reproductive Hormones During Late Anestrus.** *Once believed to be a time when most hormones were a low basal levels, more recent work has shown quite a bit of hormonal activity. It is still unsure what triggers the onset of proestrus.*
*Reprinted with permission Kal Kan Forum, Winter, 1983, Dr Patricia A. Olson.*

to fourteen months can be considered normal. It was previously believed that the reproductive organs were totally inactive during anestrus, but more recent research has shown quite a bit of activity, both in the pituitary gland and the ovaries. It is not yet known what signals the end of anestrus and leads to the next period of activity.

There is no seasonal peak of reproductive activity in dogs, except in Basenjis, which cycle once a year in the late winter, and Tibetan Mastiffs, which cycle during the autumn months. Other breeds have lost the seasonality that is seen in wild canines and may cycle at any time during the year.

# Proestrus

**Proestrus begins with the first appearance of a blood-tinged vaginal discharge and/or vulvar swelling and ends with the bitch's first acceptance of mating.**

The hypothalamus secretes gonadotropin-releasing hormone (GnRH), which stimulates the pituitary gland to produce follicle-stimulating hormone (FSH) and luteinizing hormone (LH), which, in turn, stimulate the ovary. Under the influence of FSH and LH, ovarian follicles grow and develop and are prepared for ovulation. FSH and LH also cause the follicular lining cells to produce estradiol, the female sex hormone that causes the vulva to swell and a blood-tinged discharge to be produced. Estradiol reaches maximum levels during proestrus and decreases before estrus. Many follicles regress without reaching full development, but several, depending upon the number of ova which will eventually be produced, enlarge. The estradiol produced in the follicles, along with folliculostatin, another substance produced by follicular cells, decreases the production of FSH by a negative feedback mechanism. Increased secretion of LH stimulates the production of progesterone by the follicular lining cells. Then estradiol, in conjunction with rising progesterone levels, stimulates a surge, a fifty-fold sudden increase, of LH. This surge induces ovulation, which will be discussed in more detail under "Estrus."

Before the outward signs of proestrus can be detected, there has already been considerable activity. The ovaries contain large follicles at the onset of proestrus, and blood can usually be detected microscopically in the vaginal smear, even though it may be too little to detect visually. Estradiol also begins to prepare the uterus for pregnancy, and cells lining the oviduct and the vagina undergo changes in preparation for future events. Hairlike projections called cilia grow on the cells lining the oviduct. After ovulation, movement of the cilia creates currents of fluid that carry ova to the site of fertilization at the distal end of the oviduct, and later to the uterus for implantation. Changes in vaginal lining cells are discussed in detail under "Vaginal Cytology." During proestrus, the vaginal lining is being prepared for mating. Estradiol reaches its peak concentration on about the last day of proestrus.

On the average, proestrus lasts nine days, but the length varies from one or two days to fifteen, or even more. The reason for the variation is that the outward signs, i.e., swelling and discharge, vary in individual bitches. All bitches experience the same gradual progress of hormonal and physical changes; only the outward signs vary in intensity and duration.

Males will be attracted to the bitch during proestrus, but she will not allow breeding. In fact, some males may sense that a bitch is approaching proestrus before we can be sure of it, which is even further evidence that physiologically the reproductive system is active some time before the signs of proestrus become obvious.

# Estrus

**Estrus is the period during which the bitch will allow mating.** The average duration is nine days, but, like proestrus, it can vary from one or two days to two or three weeks. Estrous behavior in excess of two weeks would be considered excessive, even though this is seen occasionally.

At the beginning of estrus, estradiol has reached its highest level and begins to decline. The follicles influenced by LH begin to undergo a change, and the lining cells are converted to luteinized cells that produce progesterone. Progesterone levels also rise slightly. The combination of estrogen plus a little progesterone causes a couple of significant changes. It makes the bitch receptive to mating, and it appears to change the character of her secretions, making her more attractive to the male. In one study, males were tested for their response to vaginal secretions from hormone treated spayed bitches. The males were interested and excited when the bitches were treated with estrogen but were markedly more excited when estrogen plus progesterone was given.

The falling estrogen level and rising progesterone level trigger the release of luteinizing hormone (LH) from the pituitary gland. LH is released in a surge over a twenty-four- to forty-eight-hour, period, and it, in turn, stimulates ovulation. The peak typically coincides with the onset of estrus, and ovulation occurs approximately two days later, on the third day of estrus in the average individual. A surge of FSH occurs at about the same time as the LH surge.

Ovulation is spontaneous in the bitch, meaning it occurs in all individuals because of the LH trigger and does not depend upon mating. Cats, on the other hand, are induced ovulators. The stimulus of mating is necessary for their LH surge, which then triggers ovulation.

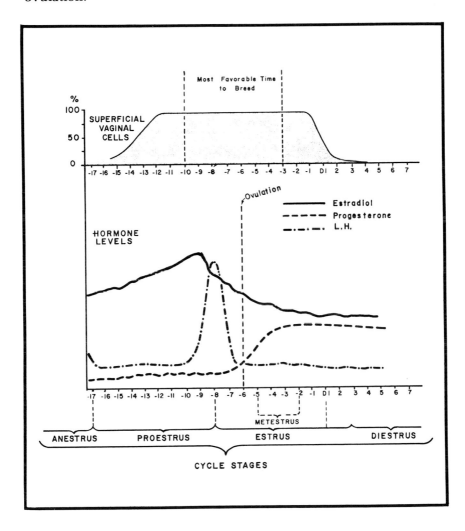

**Fig. 5.3 Hormone Levels Around the Time of Ovulation.** *The surge of LH seen at the beginning of estrus is the trigger for ovulation. After ovulation the follicles are changed to corpora lutea which produce progesterone. Important aspects of vaginal cytology are shown and are discussed in detail in chapter 7.*

After ovulation, estrogen continues to decrease and progesterone increases rapidly. It is significant that the bitch experiences estrus in this hormone situation. Females of almost all other species will be out of estrus by the time ovulation has occurred and the progesterone level increases.

Immediately following ovulation, each follicle collapses slightly, and the follicular lining cells begin to multiply and fill the vacant area. The follicular cells become luteal cells, and each follicle becomes a solid gland, the corpus luteum (CL) or the yellow body. The CLs produce progesterone, the hormone responsible for the maintenance of pregnancy.

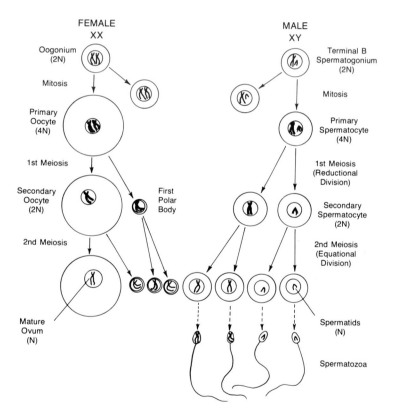

**Fig. 5.4. Meiosis** is the process whereby the diploid chromosome number (2N) of the primitive germ cells is reduced to the haploid number (N) in the mature reproductive cells. Fertilization restores the diploid number in the new conceptus. The x and y chromosomes are used in this illustration to explain how the sex of the resulting conceptus is determined only by the sex chromosome carried in the fertilizing spermatozoon.

One ovum is released from each follicle, and *because of the triggering effect of the LH surge, all follicles are ovulated during a short (twenty-four-hour) period.* In years past, it seemed perhaps logical that because a bitch would breed for many days, and because breeding at various times was fertile, the process of ovulation must be prolonged, such as the release of one egg per day or something similar. But research has put this erroneous conclusion to rest. It has now been firmly established that all the follicles ovulate within a short period of time. The number of ova released determines the maximum potential litter size and varies with the size, breed, and age of the bitch. Two to twenty or more ova may be ovulated in a given individual.

At the time of ovulation, the ova (or oocytes) are primary oocytes, immature and incapable of fertilization. Each oogonium contains the diploid chromosome number (2N). Duplication results in 4N chromosomes in the primary oocyte, and this number must be reduced by meiosis to haploid (N) before the ova are ready to be fertilized. Three days are required for maturation to a secondary oocyte, and after the ova have reached this stage, the fertilizable lifespan seems to be fairly short. The lifespan of the secondary oocyte is twenty-four- to forty-eight-hours, comparable to that of most other species, in which the ova are secondary oocytes at the time of ovulation. Thus, actual conception takes place three days after ovulation, on approximately the sixth day of estrus. The average bitch continues in estrus another three days, so it is possible, if breeding has been delayed for some reason, to breed too late. An otherwise normal, fertile individual might fail to conceive if bred late in estrus because breeding occurred beyond the fertile lifespan of the oocytes.

# Fertilization

The first meiotic division of the primary oocyte begins approximately one day after ovulation, in the middle portion of the oviduct. By the second day after ovulation, the ova have reached the distal portion of the oviduct, and the second meiotic division is taking place. About this time, regardless of when breeding occurred, the fertilizing spermatozoon enters the ovum.

Fertilization involves multiple steps in which the spermatozoon and the oocyte join to form one cell, a new individual referred to as a zygote. Each sperm cell and oocyte has been prepared by the process of meiosis to contribute one-half the normal number of chromosomes to the new individual. And it is the process of fertilization that allows for the combination of genetic material from the two parents.

38

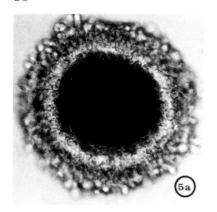

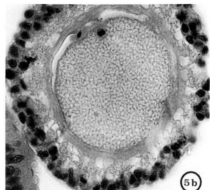

**Fig. 5.5 A.** *Beagle oocyte surrounded by follicular cells, immediately following ovulation.*
   **B.** *Thin section of a secondary oocyte with its first polar body.*
   **C.** *Thin section showing the second meiotic division in progress. The nucleus of the oocyte is arranged in a meiotic spindle at the periphery.*
   **D.** *An adjacent section of the oocyte in C. Division of the first polar body is taking place.*
   **E.** *Thin section of a fertilized oocyte. The male and female pronuclei are lying in the center of the oocyte, but have not completed fusion.*
*A,C,D reprinted by permission, Biol. Reprod. 5 (1971)*

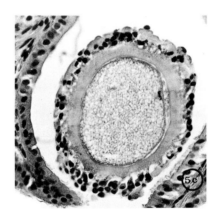

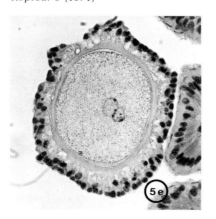

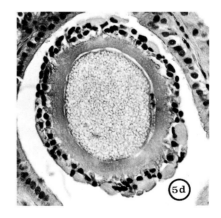

For successful fertilization to occur, one and only one spermatozoon must penetrate each oocyte. This is facilitated by enzymes on the sperm's head which allow the sperm to penetrate the egg, and by reaction of the egg's outer shell to block the entrance of any additional sperm after the first. After entrance of the sperm, the male's genetic material quickly forms a pronucleus, which then joins with the female pronucleus. The first cell division (cleavage) of the zygote follows quickly. One cell becomes two, then four, then eight. By seven days after ovulation, or four days after fertilization, or about the last day of estrus, an undifferentiated embryo of thirty-two to sixty-four cells is ready to enter the uterus.

The sequence of events beginning with the estradiol peak, and continuing through the LH surge, ovulation, fertilization, and early development of the embryo, is consistent in every bitch. But a given bitch's behavior can be quite different than average. The beginning of estrus usually coincides with the LH surge, for example, but a variation of four to five days in either direction has been observed.

# Diestrus

**Diestrus is the two-month stage following estrus when the reproductive organs are under the influence of progesterone from the corpora lutea.**

Older and, unfortunately, more recent literature calls the stage following estrus "metestrus." This term was used for many years to describe the bitch's cycle, whereas "diestrus" was used for all other species to describe the same stage. "Metestrus" actually refers to a short transitional period, just a few days after ovulation, when a shift occurs from predominantly estrogens (estrus) to progesterone (diestrus). Since there is no way to recognize it clinically, the brief period of metestrus is of academic interest only.

Diestrus is the period of pregnancy in a bitch that is successfully bred. If she is not pregnant, diestrus is the same length of time— about two months—and may be manifested as varying degrees of pseudopregnancy. Hormone levels are the same for both pregnant and nonpregnant individuals, so no hormonal assay can be used to detect pregnancy. Progesterone reaches maximum levels around the fifteenth day of diestrus and then decreases gradually.

The beginning of diestrus may be defined as the first refusal of the bitch, but considering the wide variation seen in behavior, I prefer to define it as the day on which the vaginal smear shows a shift from mostly cornified to mostly noncornified epithelial cells. This defini-

40

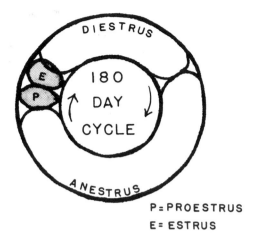

P = PROESTRUS
E = ESTRUS

Fig. 5.6 The Stages of the Estrous Cycle. *We divided the continuous cycle into separate stages to facilitate our understanding and communication. The shaded areas represent the stages during which a bitch is "in heat". The length of anestrus varies, while the length of the other stages is consistent.*

tion will be used throughout the book, even though the use of behavior to define diestrus has some merit. This change in vaginal cytology is closely related to levels of estradiol and progesterone and correlates closely with ovulation, embryonic development, and gestation length. First refusal of the male occurs at about the time of the beginning of diestrus but is not nearly as well correlated. On the average, a bitch will first refuse breeding on the second to third day of diestrus. Thus, there is an overlap between the end of estrus and the beginning of diestrus. A bitch may still breed during the first few days of diestrus, or she may begin refusal before diestrus begins. The situation occurs because estrus is defined solely by behavior and diestrus by a change in the vaginal cytology, another instance to reinforce the concept that the cycle is continuous, orderly, and unified, even though we arbitrarily try to separate it into stages.

Following diestrus, the bitch gradually returns to anestrus. What causes the corpora lutea to finally regress and progesterone secretion to cease is unknown. It has been reported that the uterus requires up to 150 days after ovulation to complete its involution and to be completely regenerated. This may explain why some bitches that have short cycles of say three to four months have fertility problems. The uterus may never completely recover, regenerate, and prepare for the next cycle, thus interfering with implantation or some other aspect of early pregnancy, such as secretions from the uterine glands.

Chapter Six

# EVERY BITCH HAS HER OWN STYLE: PHYSICAL ASPECTS OF THE ESTROUS CYCLE

The physical signs that we look for during the bitch's cycle can be divided into three categories: (1) swelling of the vulva, (2) vaginal discharge, and (3) response of the vulva to touch.

*Swelling of the lips of the vulva* and the tissue around the vulva is stimulated by elevated estrogens. To my knowledge, no research has investigated why some bitches have more swelling than others, or why some have more swelling during one estrous cycle than another. Is it the actual amount of estrogen secreted? The answer is not known. If you inject a spayed bitch with estrogens, her vulva will become enlarged.

Normally, swelling of the vulva is detectable at the same time as the blood-tinged vaginal discharge at the onset of proestrus. The swelling increases gradually during proestrus and reaches a maximum late in proestrus, at which time the lips of the vulva and adjacent tissue are firm, pinkish, and inflexible (turgid). At about the onset of estrus, the turgidity decreases and the vulvar area softens; the pink color of the membranes typically decreases at the same time. This softening of the vulva facilitates breeding and is useful in determining when to breed. As estrus progresses, the vulva becomes a bit more flaccid (soft, flexible), and at the onset of diestrus, it is usually possible to see a decrease in the size of the tissue. The lips of the vulva may actually take on a greyish cast at the beginning of diestrus.

Following a bitch's first estrous cycle, the vulva regresses, but usually not to its original size. Only if she is later spayed will her vulva return to its original prepubertal size. So it is usually possible to examine a young bitch's vulva and determine whether she has ever had a cycle.

The decrease to minimal anestral size of the vulva is gradual, and during the first month following mating there is probably no difference between a bitch that is pregnant and one that is not. But after a month or so, the vulva of the pregnant bitch remains a little larger and is more flaccid and relaxed compared to a nonpregnant bitch. This is not a primary nor especially reliable sign of pregnancy, but it can be used as a secondary supportive sign.

*The vaginal discharge* has always been used to recognize and define the stages of the bitch's cycle. Unfortunately, many of the descriptions have been greatly oversimplified, leading to confusion and misunderstanding.

A discharge normally begins at the onset of proestrus, or rather, proestrus begins when you first see a discharge. Blood from the uterus is passed down the tract and eliminated through the vagina. It is not the same at all as the flow seen during menstruation in women. Menstruation is a sloughing of the uterine lining following ovulation and a failure to become pregnant. Estrual bleeding is caused by leakage of blood through the intact lining of the uterus during the preparatory

**Fig. 6.1 The Vulva of a Prepubertal Bitch** *is relatively small and inconspicuous.*

stage before ovulation. Estrogen stimulates the uterus and increases its size and blood supply, and blood cells are lost *per diapedesis* into the lumen. This means that the blood cells simply migrate out of vessels and through tissue. No vessels are broken down in the process. The amount of blood discharged varies tremendously among individual bitches. What might be considered average at the onset of proestrus is enough blood to make the discharge bright red. Many bitches, however, have fewer cells, and the discharge will be light red, pink, amber, or even colorless. Each of these variations is normal. If vaginal smears are checked before proestrus, it is usually possible to see red blood cells several days before there are enough to make a visibly bloody discharge.

The amount of discharge is as variable as the color. The bitch tends to have less discharge as she ages, and some breeds or families tend to have more or less, which makes it appear that the trait is inherited, like most aspects of physiology.

The color of the discharge generally lightens as proestrus advances. The lightening of the color is simply a decrease in the number

**Fig. 6.2 The Vulva of a Bitch in Estrus** *is normally markedly swollen compared with its condition in anestrus.*

of red blood cells. Diluted blood is light red to pink to amber, depending upon the dilution. The most common mistake is to rely on the color of the discharge in making a decision as to when to breed—most frequently to wait for a "straw" color before breeding. In some bitches, the color will never turn pale, and in others it will be pale from the first day. You must not make the mistake of using color or any one sign to determine a breeding program. It is simply poor management.

At the end of estrus, there are several normal variations in the nature of the discharge. One involves the bitch that sheds a lot of blood and has a bright red discharge in proestrus, retaining a red color throughout estrus. As her season ends, she may still be passing blood cells in large numbers, but the color will change to dark brown. This color is due to old, oxidized blood.

A bitch that has had little or no color to the discharge during proestrus may simply stop without any noticeable preceding change. A more average bitch whose discharge changes from bright red to light pink or amber may have a colorless discharge before finishing her cycle.

At the end of estrus, the vaginal epithelial cells and white blood cells influence the appearance of the discharge along with the red cells. Immediately before diestrus begins, the vaginal epithelium appears to be sloughed, especially in some individuals, and large sheets of cells are seen for a day or two. At the same time, white blood cells reappear in the vagina, sometimes in tremendous numbers. Both these cell types if present in large enough numbers, will impart a thick texture and a white or creamy color to the discharge.

Imagine now the various colors and consistencies that may appear when diestrus begins. Various amounts of red cells will be mixed with various amounts of epithelial and white cells. Thus, the discharge may be thin, watery, and clear to thick and creamy, muddy brown in color to pinkish, amber, or colorless. Many variations are normal.

During the early part of diestrus, some discharge may persist, especially in bitches that pass a lot of blood during estrus. A small amount of brownish mucoid discharge is not abnormal. Normal discharge will be almost odorless. Signs of infection include large amounts of yellowish or greenish discharge, or discharge with an odor. Irritation or inflammation of the vulva is also a sign of trouble.

The vulva and adjacent area become sensitive to touch during estrus, and this sensitivity can be used as a sign of readiness to breed. During proestrus, if the skin above the vulva and below the anus is touched, there will probably be no response. As the cycle progresses, this area will become responsive, and when stimulated during estrus, the bitch will flag her tail and the vulvar area will be elevated to the position needed during mating. The anus will also contract when the area is stimulated. This sign can be used as an indication of readiness to breed when a male is not available for teasing.

**Fig. 6.3 The Vulva is Elevated and the Tail Held to the Side (Flagging)** *when the area is touched during estrus.*

## Chapter Seven

# WHERE TO LOOK WHEN YOUR EYES CAN'T SEE ENOUGH: VAGINAL CYTOLOGY

The examination of cells exfoliated (sloughed) from the wall of the vagina is the most useful tool we have in understanding and managing breeding in dogs. When we are confused by the bitch's behavior, in doubt about the significance of her discharge, wonder whether she is really in heat because of a marginal amount of vulvar swelling, we need not stay in the dark. All we have to do is switch on the light—the light of our microscope.

The principle of cytology is simple. The lining (epithelium) of the vagina is sensitive to levels of estradiol and it changes during the estrous cycle. The number of layers of cells increases as estrogen levels increase. During anestrus, the epithelium consists of two or three layers of small cuboidal cells. Early in proestrus, probably before proestrus is detectable externally, the number of cell layers increases, and the epithelium becomes a fully developed, stratified squamous epithelium that is twenty to thirty cells thick. Stratified means in layers, and squamous means made up of flat, large, protective cells. The body is covered with skin and the mouth lined with mucous membrane, both examples of a stratified squamous epithelium. The function in the vagina, the mouth, or the body's surface is the same: mechanical protection.

When we examine cells from the vagina, we are taking a look at the cells that are being exfoliated from the surface at that particular time. For sure, the hormonal changes that are responsible for the appearance of the cells we see may have occurred days before, but this does not detract from the usefulness. The changes that we see can be used to determine whether the heat is normal, to help determine the proper breeding time, to diagnose infection, and to prove or disprove a misalliance.

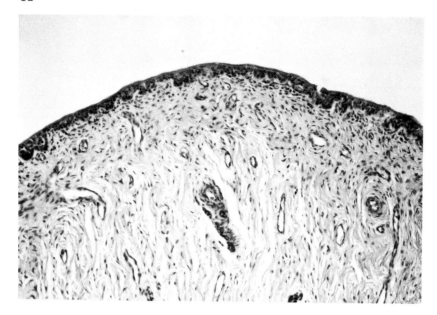

**Fig. 7.1 The Vaginal Epithelium During Anesturs** *consists of two to three layers of cuboidal cells.*

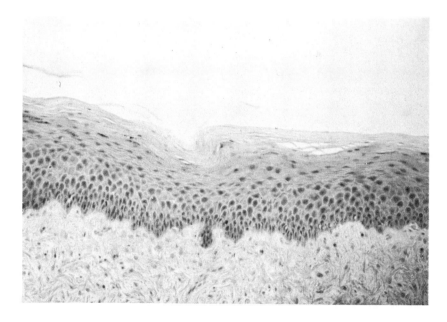

**Fig. 7.2 The Vaginal Epithelium During Estrus** *consits of 20-30 layers of stratified squamous epithelial cells.*

# Preparation of Vaginal Smears

## Materials

The materials needed to make smears are:
1. Glass microscope slides,
2. Blunt-ended glass pipette with rubber bulb, or sterile, cotton-tipped swabs, or smooth glass rod, spatula, scalpel handle,
3. Isotonic saline solution (PSS),
4. Stain,
5. Microscope.

## Procedure

1. The slides must be labeled. Ideally, use frosted-end slides. You can write on these with an ordinary lead pencil. Plain slides can be labeled by covering one end with cellophane or masking tape and writing with pencil on the tape, or by use of a diamond pen.

   The bitch's name and the full date (day, month, year) should be included. Other helpful information may be included, such as day of heat, bred that day, etc.

Fig. 7.3 **The Equipment Needed to Prepare Vaginal Smears** includes microscope slides, a pencil to label the slides, isotonic saline solution, a glass pipette, fixative, and stain.

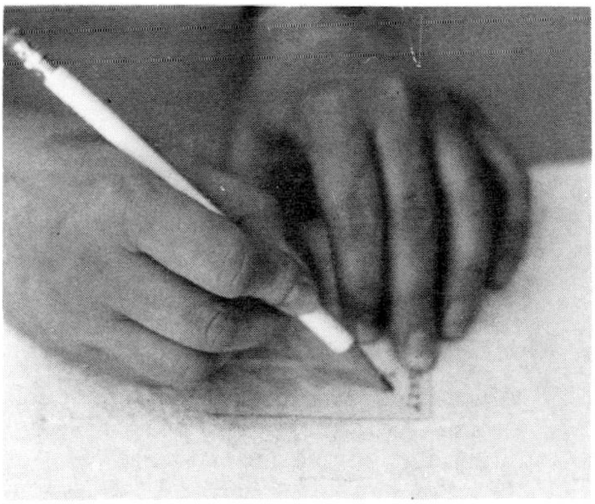

Fig. 7.4 **Labeling Each Slide is Important.** Proper labels identify every smear and prevent mixups at a later time.

2. To collect the vaginal cells, my favorite method is to use a blunt glass pipette. A small column of saline is drawn up into the pipette and then slipped through the lips of the vulva into the caudal vagina. Ideally, the bitch should be lying on her side. This way, the pipette does not have to be tipped upside down, which usually results in all of the fluid flowing back up into the rubber bulb.

Fig. 7.5 A Column of Saline Solution *is drawn into the pipette.*

Fig. 7.6 The Pipette is Inserted Gently into the Caudal Vagina. *The bitch should be lying down to prevent the fluid from flowing back into the bulb of the pipette.*

Fig. 7.7 The Pipette is Inserted its Full Length *into the vagina and the bulb squeezed gently, only partially, several times.*

Insert the pipette to its full length. Squeeze the bulb very gently several times. The goal is not to squeeze all the fluid from the pipette, but to make the fluid column wash rapidly back and forth and in so doing pick up a sample of cells.

3. Withdraw the pipette. Most of the fluid may have been lost, but even a single small drop is enough to make an excellent smear. Next, place a small drop of your sample near the labeled end of the slide, holding the slide in a semi-vertical position. Allow the drop to immediately run down the length of the slide, and blot the excess fluid off the end by touching the slide to a paper towel or tissue on the table surface. Allow the slide to dry propped vertically against something on the table. Quick drying may be hastened by holding the slide securely in your hand by the labeled end and waving it through the air.

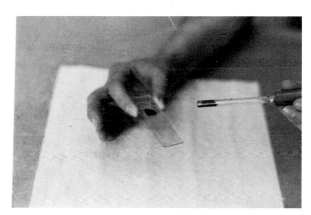

**Fig. 7.8 A Single Drop of the Vaginal Fluid is Placed on the Slide** near the labeled end. (Stain is used here to make the fluid more visible.)

**Fig. 7.9 The Drop is Allowed to Run Down the Length** of the slide and the excess is blotted from the end. The slide is rested vertically to air dry.

Another acceptable method is to use a cotton-tipped swab instead of a pipette. The swab should be moistened with saline before inserting to lessen the risk of irritating the wall of the vagina. Some kind of speculum should be used to make a channel through the vulva for inserting the swab. If the swab is inserted as is through the lips of the vulva, it will pick up cells on entry and exit. We are not really interested in examining the cells from the vulva and vestibule, but only from the vagina proper.

After inserting the swab its full length through the speculum, twist it approximately one full turn to pick up cells, then withdraw it. To apply the sample to the slide, place the cotton tip on the slide, lying flat on the table, and roll it across the surface.

This method is slightly more uncomfortable for the bitch and may have the disadvantage of actually rubbing the wall of the vagina and thereby picking up cells from deeper in the epithelium. That could confuse our interpretation, which is based upon examination of the cells recently exfoliated.

Another method less commonly used involves the use of some solid, smooth rod or elongated object, such as a scalpel handle or glass stirring rod. This would be inserted gently into the vagina, then turned and withdrawn, and the surface smeared onto the slide. A speculum should be used with this technique just as with the swab to avoid picking up vulvar or vestibular cells.

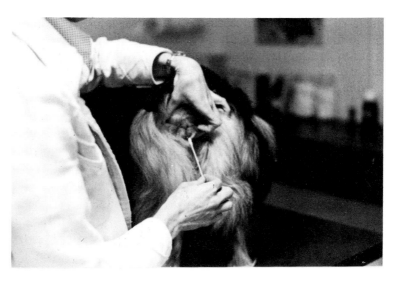

**Fig. 7.10 Vaginal Smears May Also be Collected Using a Sterile Cotton Swab.** *The swab should be moistened with saline solution and ideally a speculum should be used to avoid contact with the vulva.*

# Staining Techniques

Many staining procedures are excellent for examination of a vaginal smear. The goal here is to stain the cells so that they can be recognized by type and stage of maturity. The Paps stain used in human vaginal/cervical cytology is not recommended because it is time consuming and cumbersome. The differential staining features of the Paps smear do not really help in evaluating a canine smear. All we need to do is see the nucleus and the cytoplasm of the cells. Any stain used for blood smears will do.

## New Methylene Blue

This is usually considered a temporary stain, but this is not necessarily so. It can be used to stain the smear either wet or dried. If you are in a hurry, mix a drop of vaginal sample with a drop of stain, cover it with a glass coverslip, and examine it. If you have a dried smear, put a drop of stain on the coverslip and place it on the slide. I have discovered that the smear can be saved and is of fairly good quality if you remove the coverslip while the stain is still wet and allow the slide to dry. The cells stain blue.

## Giemsa Stain

Giemsa stain is a variation of the Wright's blood stain, is simple to use, and gives a very pretty permanent smear. The stain is purchased in a concentrated stock solution. To use, mix one drop of stain with 1 ml ( = 1 cc) of distilled water. Either have the dilute working solution in a jar of some kind and place the smear in it, or lie the smear flat on the table and pipette the stain onto the slide. Allow it to stain for fifteen minutes. Rinse it in tap water, and allow it to dry. The cytoplasm stains blue, the nuclei purplish, the white blood cell nuclei dark blue, and the red blood cells greyish blue to pink, depending upon the pH of your stain solution. Your diluted solution can be used to stain several slides or batches of slides before it must be discarded.

## Wright's Stain

This stain gives results similar to the Giemsa stain and provides a nice permanent slide. It comes in various quick-staining versions, the best of which is probably the *Diff-Quik®* staining kit made by Harleco Drug Company.

# Interpretation of the Smears

## Cells Seen in the Smear

**Epithelial Cells.** Parabasal cells are the least mature epithelial cells likely to be found in the smear. They are the ones just one or two layers removed from the vaginal basal cells. Basal cells are never sloughed but are the source of all the other cells by mitosis. Parabasal cells are seen when the epithelium is low—in anestrus, early proestrus, and again in diestrus. Their numbers are never high, and in some individuals you may never see a true parabasal cell.

Intermediate cells come in a wide range of sizes and types because they are everything between parabasal and the fully mature superficial cells. The less mature intermediate cells are small, with a relatively large nucleus. At first they are round; later in their maturation they are angular. The size of the nucleus does not really change, but the cell enlarges and flattens so that it appears smaller in relation to the

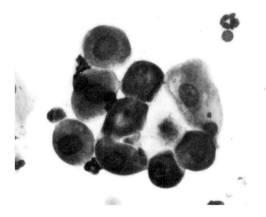

**Fig. 7.11 Small Intermediate Cells.** *These cells are exfoliated when the vaginal lining is low, in proestrus, diestrus and anestrus.*

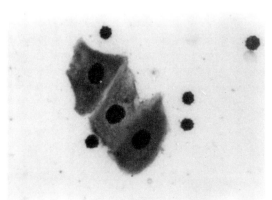

**Fig. 7. 12 Intermediate Cells** *and red blood cells.*

cytoplasm in the more mature intermediate cells. These cells are often shed in clusters. A superficial intermediate or large intermediate cell is one that has a mature squamous development of the cytoplasm and a nucleus that is readily visible and not yet pyknotic.

Superficial or cornified cells are the final, fully mature cells that line the vagina during most of estrus. These cells have keratin incorporated into the cytoplasm. Keratin is a protein substance also found in skin cells and elsewhere. Its function is protection. The term "cornified" refers to cells that contain keratin and derives from the same root as "horn" or "horny."

Superficial cells are flat and irregular in shape, usually with several angles and several relatively straight sides. The nucleus is either absent or barely visible as a shadow, or it may appear small, dark, and shrunken, (pyknotic). Whether the nucleus is visible is immaterial to whether a cell is superficial. What makes it superficial is that it is the end stage of maturation. Superficial cells are dead cells, incapable of further change.

Fig. 7.13 A More Mature Intermediate Cell, red blood cells, and white cell (arrow).

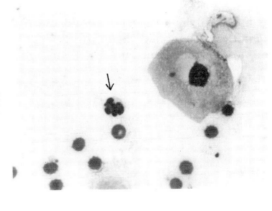

Fig. 7.14 One Large (or Superficial) Intermediate Cell, on the Left, and Three Superficial Cells. Some large bacteria are also present (arrow).

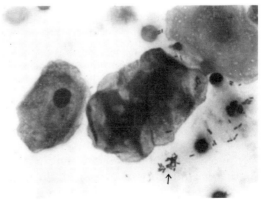

**White Blood Cells.** During anestrus, early proestrus, and again in diestrus, white blood cells, almost exclusively neutrophils, are seen in the smear. They appear to leave the bloodstream, migrate through the vaginal epithelium, and are then released free in the lumen. During later proestrus and estrus, the many layers of the vaginal epithelium block the migration of the white cells, and they accumulate in the sub-epithelial layers of the vagina. With the breakdown or sloughing of the cornified epithelium late in estrus, white cells will again appear in the lumen and thus on the smear. You will see variable numbers of white cells in proestrus, declining to none during most of estrus, and reappearing at the end of estrus or early in diestrus. In some bitches, the numbers are so great that at first glance it may appear to be an infection. The first time I saw such a smear, I was absolutely convinced that it was a terrible case of vaginitis. I insisted on examining the bitch and soon discovered that she was completely normal. When there is a vaginal infection, the white cells will also be abundant, but they will look quite different. Instead of the healthy, normal appearance, they will be degenerate, ragged, and full of bacteria and other debris. The nucleus will be less compact and stained a lighter blue color and the cytoplasm often smudged.

**Red Blood Cells.** Red cells are usually present during proestrus, estrus, and early diestrus. They can actually be seen microscopically in many cases before any color is visible to the eye. The numbers vary proportionately with the color of the discharge, i.e., when the discharge is bright red there are many red cells and when it is pale or colorless there will be few or none. Generally, I ignore the red cells in the smear. There are so many normal patterns, from virtually no red cells to huge numbers throughout proestrus, estrus, and early diestrus, that no correlation can be made with the other more important patterns seen in the vaginal epithelial cells. What many people refer to as a "silent heat" is really nothing but a normal heat with so few red cells that color is not possible to detect in the vaginal discharge.

Most of the red cells come from the uterus by a process called diapedesis. Estrogen increases the blood supply and size of the vessels in the lining of the uterus. Red cells leave the vessels and travel through the tissue in a way very similar to that of the white blood cells in the vagina. They get into the lumen of the uterus and pass through the cervix into the vagina and are then expelled through the vulva. This type of bleeding does not involve the breakdown of any blood vessels, nor of any tissue, and so is entirely different than the bleeding seen during menstruation in primates, including humans.

**Bacteria.** The vagina is not a sterile chamber, and it is normal to see various kinds of bacteria. The number varies but usually is fairly low during anestrus and diestrus and highest during proestrus and estrus. Overwhelming numbers of bacteria and many reactive degenerating white cells are abnormal and suggest a vaginal infection.

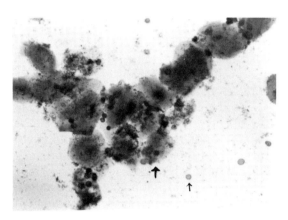

**Fig 7.15 A Fully Cornified Smear Which Contains a High Bacterial Population.** *A few red blood cells (small arrow) and large intermediate cells (large arrow) are seen.*

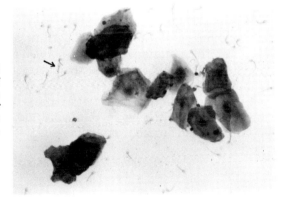

**Fig 7.16 A Fully Cornified Smear Relatively Free of Bacteria.** *Many sperm cells are seen (arrow). This smear was made within two hours after breeding.*

## Patterns

The simplest way to understand the patterns of vaginal cytology during the stages of the estrous cycle is to do a count of the parabasal, intermediate, and superficial cells and plot them as a percentage of the whole. Figure 17 shows a daily differential count for a normal, average bitch. The percentage of parabasal and small intermediate cells versus superficial intermediate and superficial cells is shown.

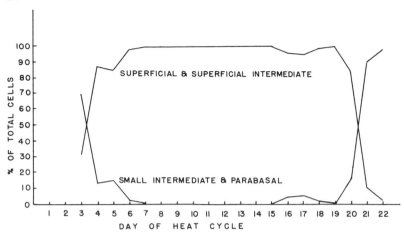

Fig. 7.17 ABOVE Daily Differential Count of Vaginal Cells *in an eighteen-month old Sheltie. This bitch was not bred, but estrous behavior (flagging) started on the eighth day and continued through the eighteenth day. Diestrus began on the twentieth day. A fully cornified smear was seen for fourteen days (day 6 through day 19).*

Fig. 7.18 BELOW Smears From a Sheltie Bitch From Early Proestrus Until the Beginning of Diestrus.

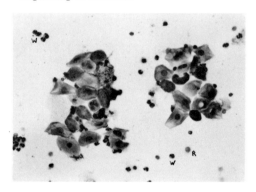

A. **Early Proestrus.** *Most of the cells are immature intermediate types. Red cells (R) and white cells (W) are also seen.*

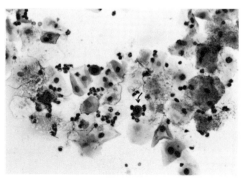

B. **Proestrus.** *More mature intermediate cells are present, along with some superficial cells. A single parabasal cell is seen (arrow).*

C. **Later in Proestrus.** *The majority of cells are now superficial. Few small intermediate cells are left.*

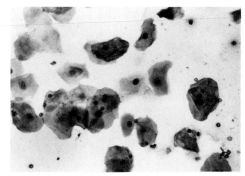

D. **A Fully Cornified Smear,** *which is normally seen from late proestrus and throughout estrus. Two large intermediate cells are present, but in many bitches they may not be seen during this stage.* **Even a Single Sperm Cell is Proof That Breeding Has Occurred.**

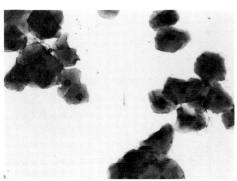

E. **The Last Day of Cornification.** *The cells are seen in clusters and large sheets. A few more immature cells and white cells (arrow) are seen.*

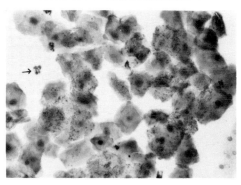

F. **The Next Day a Sudden Return of Immature Cells Signals the Onset of Diestrus.** *White cells in large numbers also reappear.*

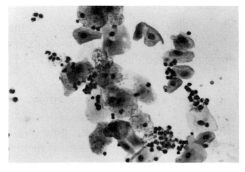

During proestrus, the percentage of superficial cells rises until it is nearly 100 percent. The percentage plateaus for ten to fourteen days, changing very little. After the ten to fourteen days, the number of superficial cells drops dramatically in a single day and continues dropping to a low count, approaching zero, within forty-eight hours. This very simple pattern—a rise in superficial cells, a plateau, and a sudden drop—is seen in bitches of all sizes, breeds, and ages that are having a normal heat.

The technique of making the smear will influence the percentage of superficial cells seen during the plateau. With the pipette and saline method it will approach 100 percent and no small intermediate cells will be seen. Using a swab, you may never see as high a percentage, and some intermediate cells may persist. So it is very important to use one technique, become familiar with the smears you get, and learn to evaluate them.

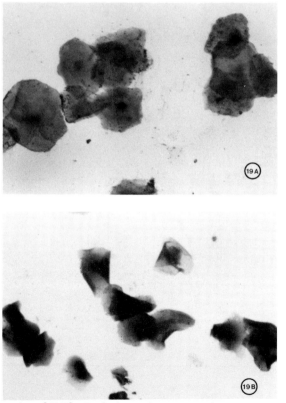

**Fig. 7.19 A and B The Technique of Preparing the Smear can Influence its Final Appearance. A.** *was prepared using saline and a pipette. The cells lie flat.* **B.** *was made with a cotton swab. The cells tend to be folded. Both smears are from the same bitch on the same day.*

The sudden decline in superficial cells observed late in estrus is the most helpful, easily observed event in the cycle. Knowing this day, we can estimate with great accuracy when the bitch ovulated, when conception occurred, the stage of development of her conceptuses, and when she is due to whelp. The reason for the close correlation is that the change is hormonally controlled and occurs at a constant time following ovulation. At ovulation, or to be more precise, approximately forty-eight hours before ovulation, estrogen levels have reached a peak and begin to decline. The stratified squamous vaginal epithelium is estrogen dependent and returns to a resting condition after estrogen levels fall. Why the change is so sudden, I cannot explain. In examining smears, it seems as though the day before diestrus, sheets of cells are seen in the smear as if the squamous layers of cells are suddenly sloughed. The cells are in various stages of degeneration with ragged edges and poorly staining cytoplasm in many cases. Regardless of the mechanism, the change is real and easy to observe.

Knowing the onset of diestrus, as determined by the smears, we can put together a time schedule of events in the cycle from before ovulation to whelping. The terminology is as follows:

$D_1$ is the day on which a sharp decline in superficial cells is seen, along with a rise in small intermediate cells. A decrease of superficial cells by at least 20 percent and a rise in the smallest intermediate cells by at least 5 percent, are the criteria for defining the event. $D_1$ is the first day of diestrus by definition.

The days preceding $D_1$ are labeled consecutively—D-1, D-2, etc., meaning the number of days before the beginning of diestrus.

The days following $D_1$ are designated $D_2$, $D_3$, etc., meaning the day of diestrus at the time being considered.

A few summary comments concerning the use of vaginal smears are in order. I have often heard smears referred to as "ovulation smears" and have heard people say that a vet looked at a smear and told the bitch's owner that the bitch was ovulating that day or that she would ovulate in four days, or that she must be bred within the next forty-eight hours, etc. *The smears cannot be used to pinpoint ovulation, until after the fact.* During the heat cycle, the smear reaches a plateau of cornification and does not change for an average of twelve days. On the day of ovulation, the smear has been cornified, on the average, for six days and will continue to be so for another five days. No one can look at an isolated smear during estrus and know just when the bitch is going to ovulate. Happily, as will be discussed in Breeding Management, you do not really need to know exactly when she ovulates in order to breed successfully.

*Table 1. Time schedule for events in the bitch's cycle related to $D_1$*

| | |
|---|---|
| D-17 | Average first day of proestrus |
| D-12 | Average first day smear is fully cornified. |
| D-9 | Peak level of estradiol observed, followed by a decline that triggers on |
| D-8 | Surge of LH, which in turn stimulates ovulation two days later. Average first day of estrus. |
| D-6 | Ovulation. Following are three days of maturation of the ova to prepare for fertilization. |
| D-3 | Fertilization. |
| D-2 | Cleavage to two-cell embryo stage. |
| D-1 | Four-cell stage. |
| $D_1$ | Eight-cell stage. |
| $D_2$-$D_4$ | Morula stage. First refusal of the male usually occurs between $D_1$ and $D_4$. |
| $D_4$ | Morula enters uterus and soon becomes a free-floating blastocyst. |
| $D_{12}$ | The fully developed blastocyst begins attachment. |
| $D_{13}$-$D_{18}$ | Early stages of major organ formation; the individual loculi are enlarging in the uterus but are too small to palpate externally. |
| $D_{19}$ | Loculi have reached marble size and can usually be palpated if the bitch is not too large, too fat, or tense. |
| $D_{20}$-$D_{28}$ | Palpation is easiest during this time. Completion of major organ formation and beginning of the fetal stage. |
| $D_{28}$-$D_{55}$ | Fetal stage, growth, maturation. Individual loculi difficult to palpate. |
| $D_{56}$-$D_{58}$ | Whelping is most likely to occur during this time. Eighty-five percent of the bitches will whelp now, and most of those whelping either early or late have unusually large or small litters for the breed. |

The smears, likewise, cannot be used to pinpoint a specific day when a bitch ought to be bred, since there is an eight-day fertile period in every bitch's cycle (see Breeding Management). It is true that the bitch must be bred when the smear is fully cornified. Breeding before cornification is complete (an unusual situation, but which can be seen in bitches that are receptive unusually early) and in diestrus, even though the bitch may still be receptive, will not be successful.

# Misalliance

Examination of a smear following a suspected but uncertain mis-mating can be so helpful that it should be considered mandatory before any action is taken to abort. The first thing to determine is whether the bitch is in estrus, or at least whether her smear is cornified at the time. If not, there is no need to worry about a fertile mating having occurred. The second thing to check is for the presence of spermatozoa. They can be seen on a wet smear and will be motile for several hours after a mating. On a stained smear they can be seen under reduced illumination. The numbers decline progressively as time goes by, and it is normal to see a few twenty-four to forty-eight hours after breeding. During diestrus, however, sperm survival in the vagina seems to be poor, and after a known mating they may not be seen in the smear after a relatively short time. Knowing that a misalliance took place during diestrus (the most common time for receptivity beside estrus), you can avoid giving an injection of hormones for abortion, yet feel confident that an unwanted litter will not be produced.

**64**

Chapter Eight

# SEX AND THE SINGLE DOG: BREEDING BEHAVIOR

## Breeding Behavior of the Bitch

Normally a bitch in anestrus will not display any sexually oriented behavior. Young bitch puppies will occasionally mount other puppies during play and when mounted usually stand rather passively. It would not be unusual for an anestrous bitch to sniff or tease or mount a bitch in estrus, and the one that is in heat would probably respond to her similarly to the way she would respond to a male. In other words, she would snap or passively avoid or stand, depending upon the stage of her cycle.

There are often signs of a bitch approaching proestrus, sometimes several weeks before her cycle actually begins. She may tease or mount other dogs, males or females. There may be some subtle or even not so subtle changes in her appetite or attitude, energy level, or personality. To a watchful owner, these signs can be a fairly reliable signal of an approaching heat. Unfortunately, the length of time varies so much that you cannot really know when she will be in heat.

During proestrus, the typical bitch will be rather passive and not especially interested in mating nor impressed by the male's interest. Her response to his sniffing will vary. Some bitches will be extremely aggressive, snapping, and growling, while some will simply sit or lie down, turn, or walk away. As proestrus progresses, the bitch gradually becomes more interested in the male. She will sniff him, allow him to sniff her, and sometimes wrestle or initiate a chase.

At the onset of estrus, which is usually about the time estrogen levels begin to drop and progesterone begins to increase, the bitch will show true estrous behavior. She will sniff the male and invite him to sniff her by turning her rear quarters toward him. She will usually elevate and deviate her tail to one side (flag) and even tease and try to get the male to pay attention to her. She may poke him in the side with her nose or put her head across his back or a paw on his back. A truly receptive bitch that is not getting the desired response from the male may mount him, almost as if to show him what she expects.

When a bitch in estrus is mounted she will stand quietly, sometimes with rear feet placed wide, flag her tail, and allow the male to breed her. It would be considered normal for her to stand quietly, elevate her vulva to put it in a more favorable position for intromission (insertion of the penis), and continue to stand quietly during intromission until the tie and dismount are accomplished. Unfortunately, not all bitches are ideally steady, and some are prone to have a change of heart at the last minute. The vulva may be sensitive in maiden bitches or when they are at the beginning of estrus, and this will make some bitches turn, sit down, cry out, and try to escape. Some will also show signs of discomfort when the bulb of the penis swells to form the tie and try to escape at that time. But even those bitches that struggle during intromission or swelling of the bulb will normally stand fairly quietly during the tie. It would not be unusual, however, for a bitch to try to walk away or go into a doghouse, or some such behavior during the tie. The male would have no choice but to follow along with her and it would be extremely unusual for either partner to be injured. Occasionally the bitch or the dog will seem fatigued during the tie and will lie down, pulling the other one along. This, likewise, is unlikely to do any harm.

After the tie, the bitch will sit down and lick her vulva. There is usually a discharge of semen at that time, and apparently the bitch senses the need to clean herself. The bitch's activity following breeding will have no effect on conception. Whether she is allowed to urinate or not, whether she is kept quiet or exercised, will not matter. Any sperm that will participate in conception are in the uterus and oviducts and safe from loss by the time the tie is broken.

# Breeding Behavior of the Male

One of the most striking behaviors among dogs is the incredible craziness that overcomes a male when he comes in contact with a female in heat. His attention is completely focused on pursuing her, and if restrained from doing so, he will express his grief in ways that will break your heart. He may refuse to eat; he may howl to the point of being truly ultrasonic. He may disappear from home, if given the opportunity, for days at a time. He will leap tall fences in a single bound! He may open doors and break through windows that seemed indestructible.

It has been reported that the scent that stimulates the male—the sex pheromones present in the bitch's secretions and urine—is detectable over extremely long distances. I used to believe that myself, and whenever I had a bitch in heat, a pack of ardent suitors would gather around the house, making it impossible even to let the bitch exercise inside the fenced yard without constant supervision. At that time, though, I had the habit of taking my dogs for a walk almost every evening after supper. We walked for many blocks in every direction, and as I now realize, the bitch left a scent trail for the males to follow straight back to our door. Of course, any dog with a sniffer worth its salt could find her. When this situation occurred to me, I changed the nightly ritual. I carried the bitch to the car, drove away about four blocks to a local park, unloaded, and commenced the walk from there. From the first time I used this routine, not a single dog discovered my bitches in heat, while they were confined inside my fence. Apparently even that much physical barrier was enough to prevent detection by the local wandering Romeos. I am sure that if one had come to the fence at a time when the bitch was out and investigated nose-to-nose, it would have been enough to let out her secret.

Young male puppies sometimes demonstrate sexually oriented behavior when only four or five weeks old, mounting littermates and thrusting. This behavior is normal and natural. The physical contact with other puppies in play is important in the development of normal behavior. Pups raised in isolation from other dogs often have problems when they are later expected to breed. They may mount, but the mounts may not be properly oriented. They also may direct their sexual interest toward their owner rather than toward another dog. In some cases, if the male is extremely attached to and protective of his owner, he may regard a bitch in estrus as in intruder and aggressively attack her. It can be detrimental to a dog's sexual performance if his early sexual interests are discouraged. In other words, don't scold a young dog for mounting other dogs. In fact, it would be better to encourage it, although in a moderate way. After all, even the best stud dog can be expected to have some manners. Scent marking with urine inside the house does not need to be tolerated, either, even though it is a natural response of a dog when sexually excited. He can learn some limits, and inside the house should be considered off limits in any case.

When a sexually mature male encounters a female in estrus, he will exhibit a series of behavior patterns that eventually lead to copulation (breeding). The behavior can be divided into several distinct individual patterns, and exactly which are exhibited and to what extent varies from individual to individual and from situation to situation. The behavior of the female also has considerable effect on the male's responses. The male's behavior may include:

- stiffly wagging tail, elevated.
- sniffing nose, ears, neck.
- sniffing flank.
- sniffing area of vulva.
- licking ears, neck, back.
- licking vulva.
- play bowing.
- wrestling.
- chasing bitch.
- urinating.
- standing parallel to bitch.
- standing with head over bitch's back.
- placing one or both paws on bitch's back.
- mounting and clasping flanks.
- pelvic thrusting.

The preliminary approach, the initial sniffing and licking, seem to serve the purpose of testing the bitch's reaction. If she is disinterested or aggressive toward his advances, the male may not try mounting as soon as if she seems either passive or willing to stand. I have never seen a male who would become involved in a fight with a snapping or growling bitch. He would simply turn away to avoid being bitten and try another approach.

When the bitch is ready for breeding, she will stand solidly, and copulation will usually be accomplished. The male clasps her around the flanks just in front of the rear legs. With the penis partially erect, he thrusts until by trial and error the tip of the penis enters the vagina between the lips of the vulva. The contact initiates a more intense involuntary thrusting as the penis continues to erect and enters the vagina. At this time, the dog usually is partially elevated over the bitch's back and dances from one side to the other on his toes with both feet at times off the ground simultaneously. This intense, deep thrusting is called the intense ejaculatory response, and it is at this time that the bulb of the penis enlarges and ejaculation begins. The enlargement of the bulb within the vagina causes the penis to be locked inside the vulva, and the lock or tie is accomplished.

When the intense ejaculatory response is complete and the bulb is enlarged, the dog will dismount, placing both front feet on the ground on one side of the bitch. Then he lifts the rear leg on the opposite side over the bitch's back so that he can turn, and the remainder of the tie is spent with the dog and bitch standing tail to tail. The length of the tie varies from a minute or less to over an hour. During this time, ejaculation continues intermittently with small amounts of sperm-free prostatic fluid being deposited. The tip of the penis remains in its original orientation, and the 180 degree turn is accomplished by a twist in the penis just behind the bulb. The penis is extremely elastic in this area, and the twisting appears to cause no discomfort to the dog. The purpose of this particular aspect of copulation in dogs in unknown. Conception can occur without a tie, both in natural mating and in artificial insemination. The length of the tie appears to have no influence on conception or any other aspect of reproduction.

During the tie, the bitch usually will stand quietly, but some will become restless and excited and possibly try to get away. In so doing, the male will be forced to come along, dragged by the bitch. While there is probably little potential danger if either partner sits or lies down or struggles, for complete safety they probably should be held for the duration of the tie. When the tie is long (over twenty or thirty minutes), both partners seem to become very tired, and they can be laid down if they will tolerate it.

Throwing a bucket of water on a mating pair of dogs will not accomplish anything but cool them off temporarily. But there is a way to break a tie if it is ever desirable to do so. I tried this on one bitch brought to my stud dog and was able to make it work. These dogs were Shetland Sheepdogs and were therefore small, and it might not be so easy with the larger breeds. My reason for doing it was mainly curiosity, but during their first breeding, the tie lasted over an hour. That was more than any of us wanted to go through again. On the second breeding, I tried this technique after twenty minutes and was able to get them apart in about two minutes.

First, you must turn the dog back to his original position at the beginning of copulation, with forelegs up over the bitch's flanks. Then, sit behind the pair and hold them tight against your body, pushing him forward as hard as possible toward her, and pulling her back. The idea is to push his penis forward in the vagina a little bit, thus relieving the pressure behind the bulb and allowing the engorgement to reduce and the tie to be broken. Hold in this position for a minute or more.

When the swelling of the bulb of the penis goes down, the tie will be broken and the dog and bitch will quietly separate. Both will usually sit down and lick their genital area. The dog will also usually show some interest in the bitch at this time and will sniff and lick her vulvar area. Once in a while the skin at the tip of the prepuce will roll inward when the penis is withdrawn into the sheath. When

**70**

**Fig. 8 1-6 Courtship Behavior.** *These Labrador Retrievers engaged in some, although not all, of the common behaviors seen before mating. 1. Chasing 2. Lifting a foreleg in invitation to play. 3. Sniffing 4. Licking 5. Inappropriate mounting of the head 6. Inappropriate mounting of the side.*

1

2

3

4

5

6

**Fig. 8 7-12 Copulation.** *Successful mating includes: 7. Mounting and pelvic thrusting. The bitch is standing quietly. 8. Deep pelvic thrusting (involuntary) to achieve intromission 9. Beginning of dismount 10. Dismount 11. Male lifting leg over to finish dismount 12. The coital tie or lock.*

7

8

9

10

11

12

this happens, the tip of the penis may not retract entirely into the sheath and will be exposed to drying and trauma. The dog's licking usually corrects the problem, but it is still a good idea to inspect the dog's prepuce after every mating to be sure that everything is in place.

There is a refractory period after a mating until a male is willing, or able, to mate again. In a 1967 study by B.L. Hart, the following was found:

**Percent of Second Matings Within Specified Time Intervals**

| Time after mating | % males which mated within 15 minutes of introduction to second bitch |
|---|---|
| 30 minutes | 40 |
| 2 hours | 69 |
| 4 hours | 69 |
| 24 hours | 95 |
| 48 hours | 100 |

The dog's environment can have an effect upon his sexual behavior. Males tend to be more territorial than females, and their sexual behavior may be inhibited when in a strange place. Therefore, it is better to bring the bitch to the male's home for breeding.

People may have an influence on behavior, too. Some dogs will not mate when they are being watched by people, while others prefer to have a familiar person in attendance. Most dogs must gradually be made used to having the bitch held for them, and it is worth the effort required to train them this way. It is especially helpful in holding up a submissive bitch that continually sits down, as well as an aggressive one that turns, snaps, or tries to escape.

Experience is important in the male's behavior. Young males are more likely to mount inappropriately (the head, shoulder, side) and often do not dismount after the tie is formed, until they have gained some experience.

## Chapter Nine

# SUCCESS COMES WITH SOUNDNESS

## Breeding Soundness of the Bitch

A breeding soundness examination of a female may be done before her first breeding or after she has been bred and has experienced some kind of problem, the cause of which needs to be discovered.

Perhaps the most important part of the examination and by far the most helpful for a veterinarian, is the history of previous reproductive activity. Ideally every owner should keep a written record for each bitch, including the following:

1. date of first signs of proestrus.
2. nature of the vulvar swelling and discharge during the cycle.
3. bitch's behavior, when she first started to flag or flirt, whether she was teased with a male.
4. breeding dates.
5. vaginal cytology information, if any.
6. dates of first and last refusal of the male.
7. dates when palpation to determine pregnancy was done, with results.
8. whelping dates.
9. performance as a mother.
10. signs of false pregnancy, with dates.

The bitch's socialization, both to people and to other dogs, is very important. The bitch puppy should stay with her litter for at least seven weeks. She needs to learn dominance/subordination relationships and bite inhibition from her mother. It is important that she be socialized with other dogs between eight and twelve weeks of age. A puppy taken into a new home where she never has contact with other dogs may not be able to function normally when she is later expected to breed.

Both the bitch's general health and her excellence as a representative of her breed must weigh strongly in her evaluation as a breeding prospect. She must be strong, active, temperamentally sound, carry excellent coat, and be what is commonly called an "easy keeper." She should be able to maintain excellent condition when fed a good balanced diet without a lot of supplements, and she should have a good appetite. Individuals that do poorly or have a poor appetite should certainly be severely faulted as breeding prospects because they are likely to produce pups that follow in their pawprints, besides being problems themselves to manage. The bitch should be kept properly vaccinated and free of parasites, both internal and external, at all times. Regular checks for heartworm are important in areas where the disease is known to be a problem.

The pre-breeding evaluation starts with a thorough physical examination by a veterinarian. A urinalysis and blood count are indicated if there have been previous problems. A chemistry panel and thyroid screen are also done to complete the general physical exam. Hypothyroid individuals show various kinds of reproductive disorders and can usually be helped with daily supplementation of thyroid hormone. The condition probably has a hereditary basis, so you must

| Sarabande Dewberry, CD | | | | Heat Periods | | |
|---|---|---|---|---|---|---|
| # | DATE | AGE | INTERVAL | DATE BRED | WHELPED | GEST |
| 1 | 10/4/78 | 8½⁺mo | — | — | — | — |
| 2 | 4/9/79 | 14½⁺mo | 6 mo | — | — | — |
| 3 | 10/12/79 | 1yr 9mo | 6 mo | 10/27,28,29/79 | 12/25/79 | 59d: D56 |
| 4 | 6/11/80 | 2yr 5mo | 8 mo | — | — | — |
| 5 | 2/2/81 | 3yr, 1mo | 7¾ mo | 2/13-18/81 | 4/17/81 | 65d: D58 |
| 6 | 11/16/81 | 3yr 10mo | 9½mo | — | — | — |
| 7 | 7/18/82 | 4yr 6mo | 8 mo | 7/30, 8/1/82 | 9/29/82 | 60d: D57 |
| 8 | 4/26/83 | 5yr 3mo | 9¼ mo | — | — | — |
| 9 | 2/7/84 | 6yr 1mo | 9⅓ mo | 2/18/84 | | |
| 10 | | | | | | |
| 11 | | | | | | |
| 12 | | | | | | |
| 13 | | | | | | |

**Fig. 9.1 A Record of the Estrous Cycles** *for every bitch can be kept on file cards. A wealth of information is readily available and easily stored with this simple system.*

be aware of this and weigh it heavily in the decision whether to risk perpetuating such a problem. In most cases, it would be best to not breed a bitch known to be hypothyroid.

Next, an examination of the reproductive system is done. The size and position of the vulva are noted, and a digital exam of the vagina is done. The size of the vaginal opening can be determined, and the presence of strictures or fibrous vaginal bands can be detected. The size of the pelvis can be determined by either vaginal or rectal digital examination.

A canine brucellosis test should be done before each breeding. The test is important not because you expect to encounter the disease often, but because in those rare instances when it does occur, the consequences are so devastating. Brucellosis can cause various kinds of reproductive failure, including failure to conceive. There is no cure for the disease, and it is possible that it can be transmitted to humans. The rapid slide agglutination test is usually done, and it is a good, sensitive test. A negative test is reliable and in all but some extremely rare cases rules out the disease. A positive test is not as specific and indicates that further testing should be conducted, usually by a tube agglutination test and/or cultures.

A vaginal culture may be run if there is concern about the presence of infection or if the owners of stud dogs require it. The vagina is not a sterile chamber, and a variety of bacteria can be cultured from normal bitches. These bacteria include *Streptococcus* sp., *Staphylococcus* sp., *Escherichia coli*, *Proteus mirabilis*, and *Pseudomonas aeruginosa*. All of these bacteria have been reported to cause uterine disease and infertility, so the interpretation of the results of a vaginal culture is not always simple and clear-cut. Laboratories commonly report the presence of bacteria known to cause reproductive problems, while not specifically reporting others that are considered nonpathogenic. A typical report may read either "no growth," "no abnormal flora," or list a specific species of bacteria that was present and that could cause an infection. But its presence does not necessarily mean that there is an infection or that treatment is needed. I personally do not recommend a routine vaginal culture for every bitch regardless of her history. A vaginal smear during proestrus or estrus can indicate, in most cases, whether an active infection is present that warrants culture or treatment.

If it were physically possible, a culture of the uterus would be a useful procedure because the uterus is normally sterile, i.e., free of bacteria. Unfortunately, because the cervix, even during estrus, is so small, and because of its location and anatomy, it is virtually impossible to enter the uterus to collect the sample. A culture of the cranial vagina near the cervix is a compromise procedure. This procedure can only be done by a veterinarian. In order to get a proper specimen for culture, a guarded culture swab is used or a sterile speculum through which a culture swab can be inserted. The swab must not touch the lips of the vulva or the caudal vagina; if it does, the results will be quite meaningless. The culture should be done when the cervix is open, during proestrus or estrus, or immediately following whelping or abortion. Bacterial sensitivity to antibiotics should be tested to aid in choosing the proper drug for treatment.

In certain instances of specific reproductive problems, other tests can be performed. Hormone assays for estradiol or progesterone may be indicated, depending upon the problem. The only way to positively diagnose some uterine problems such as cystic endometrial hyperplasia or mucometra, is by biopsy. An exploratory surgery could be performed to physically examine the reproductive organs. Perhaps this procedure should be performed more often, as it is direct and specific and more likely to give an answer in some cases than any kind of laboratory procedures.

Melanie — 5th Heat — FEBRUARY 2, 1984
Last Heat : June 10, 1983
Interval    7 mos  23 days   ( 237 days )
Bred: June 20, 22, 24, 1983 Whelped: August 22, 1983

| JAN 29 1984 | 30 | 31 No Signs | FEB 1 Not checked | 2 Slight swelling, Bright red disch, not a lot DAY 1 | 3 A little more swollen Bright Red Begin Smears 2 | 4 Turgid. A little more swelling Bright red 3 |
|---|---|---|---|---|---|---|
| 5 Still more swollen Bright red Turgid 4 | 6 SAME 5 | 7 ~SAME maybe less +turgid 6 | 8 Color Lighter, Watery Red. otherwise same CORNIFIED 7 | 9 Flagging No change Quite swollen 8 | 10 Same Flirty 9 | 11 Softer. disch. is Pinkish 10 |
| 12 almost No Color 11 | 13 Colorless disch. Vulva same Flirty 12 | 14 Same 13 | 15 almost No discharge. Colorless Still flagging 14 | 16 No disch. thick, dark pink, picked up for smear 15 | 17 Dry, No Disch. Swelling down Still Flagging N/c Diestrus end smears 16 | 18 Same still flagging a little 17 |
| 19 Still flagging 18 | 20 out of heat 19 | 21 20 | 22 21 | 23 | 24 | 25 |

**Fig. 9.2 A Detailed Record of Each Bitch's Cycles** is kept in a calendar format. Daily notes only take a few seconds to record. The author keeps blank forms and the record sheets in a notebook. When the bitch is noticed in heat a sheet is made up with the appropriate calendar dates and days of her cycle.

Vaginal cytology is an extremely valuable tool. If it were possible to have a complete series of daily smears for the bitch, beginning as early as possible in proestrus and extending into the first few days of diestrus, a wealth of information would be at hand. The degree of vaginal cornification, the duration, the presence of inflammation, and the transition into diestrus are all important aspects of normal reproductive function. The question, "Did she have a normal heat cycle?" can only be answered when the vaginal smears are examined and then related to the other aspects of the bitch's cycle.

# Breeding Soundness of the Male

The male dog has a truly incredible capacity for reproduction. He is not a seasonal breeder, unlike deer or other wild animals, but is capable of reproduction the year round. He can potentially be used every other day without a decline in fertility, which enables certain, albeit few, individuals to sire many hundreds of litters in their lifetime. He is available to service bitches from all parts of the country, even though interstate shipping is at times frustrating and quite costly. Hopefully in the future, with the use of frozen semen, he can sire litters for a great while after his natural lifespan. His role in the breed can be quite different than that of any female. Considering all this, a primary consideration in a male's breeding soundness exam is an evaluation of his potential to contribute in a positive way to the gene pool of his breed. In other words, in evaluating a male for breeding soundness, first consider, "Is he worth it?" Does he have some outstanding feature or features that are worthy of perpetuating? Is he a truly outstanding representative of his breed? Does he have correct temperament and/or working ability? Is he robust, sound, and healthy? There should not be any compromise in the areas of health, temperament, and soundness. Considering the male's capacity for reproduction, only a few outstanding individuals should be considered.

The first consideration in breeding soundness is the individual's general health. He should first of all be what is commonly called an "easy keeper." He should be in vigorous good health, strong and active, with an excellent coat when fed a normal amount of a good-quality, balanced feed. A dog that is chronically too thin or too fat, with poor appetite, poor coat quality, or poor general condition should not be used for breeding. If some specific cause for his problems can be found and corrected, such as parasitism, then of course this must not be held against him. But the dog that is constitutionally a "poor doer" will be likely to produce his kind and should not be used for breeding.

A good physical examination, including tests for internal parasites, should be done. Vaccinations should be kept up-to-date at all times. Dental health should be checked, as infection of teeth or gums can have serious effects upon all other functions.

Special attention then must be given to the dog's reproductive system. He should be tested for canine brucellosis. His reproductive organs should be carefully examined, including the prepuce and penis, for localized infections or discharges. His testicles should be carefully palpated. They should be of good size for the breed, firm in texture, and correctly placed in the scrotum. The epididymis is easily palpated. The tail of the epididymis lies on the caudal end of the testis, and the head lies along the cranial end. The spermatic cord can be palpated from the testicle through the scrotum to the point of entering the body at the inguinal canal. The prostate should be palpated rectally to determine whether it is of correct size, shape, consistency, and location. Enlargement of the prostate is not unusual in older males and does not necessarily indicate a fertility problem.

The next step would be to evaluate the dog's semen (Table 1). Ideally, it should be collected with the use of a bitch in estrus as a teaser. In this way, some aspects of the male's behavior can also be evaluated, and generally a semen sample is easier to collect and is of better quality when the dog is excited by the presence of a bitch in heat. It is important to know whether he is normally excited by the bitch, whether he investigates, sniffs, and mounts in a normal fashion. The dog's ejaculatory response can usually be evaluated when collecting artificially. When properly stimulated, at least some of the intense ejaculatory response will be elicited.

**Table 1. Characteristics of Normal Semen.**

| | |
|---|---|
| Volume (ML) | less than 1-20 |
| Motility (%) | 30-90 |
| Total Sperm (Millions) | 100-1500 |
| % Normal Sperm | 35-97 |
| Sperm Concentration (Millions/ML) | 25-400 |
| pH | 5.5-6.5 |

The semen must be collected in a clean container held in the hand to keep it at approximately body temperature. It should be examined immediately for motility, concentration, and morphologic abnormalities.

The most common technique for evaluating semen quality is to estimate the percentage of sperm with progressive motility. A skilled and experienced observer is required. A progressively motile sperm cell moves forward in a straight line, and the head rotates a full 360 degree turn with each lash of the tail. Sperm that have movement of the tail but are not moving progressively forward, are spinning or moving in circles or erratically, are not considered progressively motile. There is a relationship between motility and fertility in most domestic animals. In the dog, the average motility is 75 percent with a range from 30 to 90 percent, observed in fertile individuals. At this time, there are no clearly established criteria as to what percentage of motility is essential to fertility in the dog. The semen sample must be kept as nearly as possible at body temperature to get an accurate estimate of motility. Lower temperatures will adversely affect motility. The equipment used for collection must be scrupulously clean.

Concentration of sperm (number/ml) is determined using a spectrophotometer, a device that measures the transmittance of light through a translucent liquid. The transmittance is then compared with a known standard to derive the concentration in the sample. The concentration of sperm per milliliter (ml) is then multiplied by the milliliters of semen to determine the total number of sperm in the sample. At the present time, an actual sperm count is seldom done in practice. Few veterinarians have a spectrophotometer. Other methods that involve counting cells in a diluted sample under the microscope are more practical and, although time-consuming, are readily available to veterinarians. There are no firmly defined standards yet for fertility in relation to sperm count. For purposes of artificial insemination of frozen semen, it is going to be extremely important to know this. If, for example, it is determined that for maximum fertility 100 million motile sperm must be present in the inseminating dose, it can be determined how many inseminating doses are present in an ejaculate. An ejaculate with 500 million motile sperm would provide five inseminating doses. A normal dog may have 250 million to more than 1 billion spermatozoa per ejaculate. The total sperm number is related to testicular size, which is related to body size. Thus, larger dogs tend to produce more sperm.

A reduced sperm concentration in the ejaculate may be indicative of testicular degeneration, senescence (old age), incomplete ejaculation, or blockage of the epididymis. Also, sperm concentration may be low in young dogs. Semen quality may be less in first ejaculates from young dogs as well as from mature dogs after a long rest from sexual activity. In evaluating such an individual, several ejaculates on successive days may need to be checked to get a true picture of a dog's semen quality (Table 2).

**Table 2. Characteristics of Semen Usually Associated With Decreased Fertility.**

1. *Motility less than 70%*
2. *Total Sperm count less than 100 million*
3. *Head and midpiece abnormalities exceeding 40%*
4. *Head, acrosome and protoplasmic droplet defects exceeding 20% (associated with lowered conception rate).*

The morphology or anatomy of the spermatozoa is also evaluated. The percentage of morphologically normal spermatozoa is related to fertility in some domestic animals, but it has not been shown for the dog. A stained slide should be prepared as quickly as possible after collecting the ejaculate, because some aspects of morphology can be affected by temperature.

This exam requires a good high-power (400x) microscope, or even better, an oil-immersion lens at 1000x. This exam supplements the motility estimate.

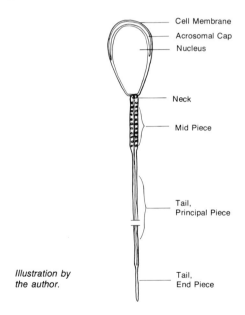

*Illustration by
the author.*

**Fig. 9.3 Anatomy of a Normal Spermatozoon.**

Abnormalities are divided into two classes: primary and secondary. Primary abnormalities are those that originate in the testes during the formation of the sperm. Secondary abnormalities are those that originate beyond the testes in the duct system, mainly in the epididymis.

Primary abnormalities involve the head, midpiece, and tail of the sperm. The head contains all the genes of the dog as DNA, plus some enzymes needed in fertilization. A high percentage of abnormalities of the head can severely reduce fertility. Heads that are too small (microcephalic), too large (macrocephalic), round, doubled, pear-shaped (pyriform), thin and elongated, or with abnormalities involving the acrosome, may be observed.

The midpiece may be kinked or doubled (along with a double tail in some cases), with a swelling somewhere along its length or attached at the edge of the head (abaxial). The tail may be doubled

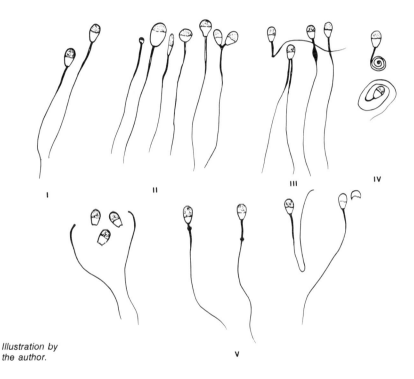

Illustration by
the author.

**Fig. 9.4 Normal and Abnormal Sperm Morphology.**
I.  Normal spermatozoa
II. Primary abnormalities of the head, left to right: Microcephalic, Macrocephalic, Elongated, Round, Pyriform, Doubled.
III. Primary abnormalities of the midpiece, left to right: Kinked, Double, Swollen, Abaxial.
IV. Primary abnormalities of the tail: Coiled.
V.  Secondary abnormalities, left to right: Detached heads and tails, Proximal protoplasmic droplet, Distal protoplasmic droplet, Bent tail, Detached acrosomal cap.

or coiled. The exact significance of the various abnormalities has never been worked out for the dog, to my knowledge. In other words, how many abnormal spermatozoa can be present before fertility is impaired is unknown. Normal dogs have been reported with 3 to 66 percent abnormalities, including both primary and secondary types.

Secondary abnormalities occur after the sperm are developed and have left the testis. The most common types of abnormalities are kinked tails and kinked midpieces, detached heads, and retained protoplasmic droplets. The location of a retained droplet can be either proximal or distal along the midpiece. A detachment of the acrosomal cap is also a secondary abnormality.

Retained protoplasmic droplets are probably the most common secondary abnormality. Every sperm cell has a small droplet of residual protoplasm as it leaves the testis, and it is normally moved distally down the midpiece toward the tail and finally lost. A high percentage of droplets, especially proximal, may indicate a dysfunction of the epididymis. Bulls with a high percentage of sperm with retained protoplasmic droplets have a reduced fertility, and the same may be true for dogs, although this has not yet been definitely established.

External factors, such as heat or cold shock, dirty equipment, and poor handling techniques, may increase the percentage of secondary abnormalities.

A final aspect of a dog's fertility soundness exam would be his willingness and ability to breed. Certain types of behavior are learned, or conditioned, by improper training and handling. A dog that has been made fearful by repeated punishment for sexual behavior may actually be quite normal. He should not be considered unsound and may be used in a breeding program even if it requires artificial insemination. On the other hand, a male with poor libido that is not the result of prior conditioning may not be a desirable sire, as the trait, like every other aspect of behavior and physiology, may be inherited by his get.

Chapter Ten

# HOW TO GET PUPPIES WHEN YOU REALLY WANT THEM: BREEDING MANAGEMENT

Management of breeding involves anything that we as outsiders do to see that a bitch in heat is bred to the male of our choosing in such a way and at such a time as to get her pregnant. The simplest management program would be to let the dog and bitch run together freely throughout her heat cycle and let nature take its course. This is the way many "backyard" and inexperienced breeders do it. But in most cases where valuable show and breeding animals are involved, that scheme is not acceptable. It is too hard on the male, too hard on the female, too uncertain, and potentially dangerous.

Any breeding management program must be aimed at a single goal: to achieve at least one mating of a fertile bitch to a fertile male during the time in her estrous cycle when she will conceive a litter. In extensive research involving Beagles, it was shown that the fertile period during estrus is centered around ovulation. A single mating anytime from four days before ovulation until three days after ovulation will result in an excellent chance of conception (almost 100 percent), assuming both the dog and the bitch are fertile (Figure 1). This fertile period is eight days in length, so our job as managers really should not be difficult. There are a number of ways to ensure a breeding at the appropriate time.

## Behavior

Most bitches will first stand for mating a few days before ovulating. On the average, first acceptance occurs two days before ovulation (Figure 2). When related to the onset of diestrus, first acceptance occurs at about D-8. The most fertile time in the cycle stretches from

D-10 to D-3, so the "average" bitch that first accepts at D-8 should have an excellent chance of conception if she is bred on her first day of acceptance.

But what about those few bitches that will accept unusually early? If we adopt a breeding program that simply allows one mating at first acceptance, a percentage of our bitches will miss because of being bred too early. In order to avoid that unfortunate event, I would advise breeding at first acceptance and then repeating the breeding approximately four days later. This is a little different from the most widely used and most traditional plan of breeding (whenever) and repeating two days later. Remembering that the goal is to achieve a mating during the bitch's fertile time, you will increase the odds of breeding a given bitch at the right time if you space the breedings a little farther apart. To breed two days apart might still result in a second breeding too early for a few bitches, but to wait four days should cover just about every possible inividual, even the earliest breeders. When circumstances allow, and the stud dog is not in heavy demand, you could do three breedings, every two, three, or four days, and have a nearly foolproof breeding management program. Do not be afraid that the bitch will conceive pups of different ages if breedings are spaced several days apart. The time of ovulation and, three days later, fertilization of the ova, determine the "age" of the conceptuses,

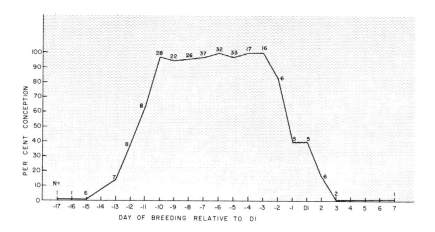

**Fig. 10.1 Conception Rate Compared With Day of Breeding** *in a study of 267 Beagles. N is the number of bitches in each group. Each bitch was bred once, and vaginal smears were used to determine D1, the onset of diestrus. The best conception rate occurred when breeding was done from D-3 to D-10, an eight day time span. Ovulation occurs on D-6 and fertilization on D-3.*

Reprinted with permission from Am.J.Vet.Res. Vol 35, No 3, 1974.

regardless of the number of matings and the number of days between the matings. Any sperm deposited before the time of conception (D-3) will survive and be available when the ova are mature.

This brings to mind another topic, somewhat peripheral, but related. If a bitch is mated to more than one male anytime before the date when conception is possible, the sperm from both, or all, will be held in storage. When conception occurs, random chance could determine which sperm fertilizes each ovum. We do not know whether sperm deposited on or near D-3 have any advantage over others. The pups in the resulting litter could be sired by more than one male. Thus, after breeding to the desired stud dog, it is extremely important to prevent mating to any other dog. It is not safe to assume that because the planned mating has occurred, the bitch is no longer in jeopardy of becoming pregnant by another unwanted dog. She may not experience fertilization until several days after the mating, depending on the timing.

A question comes immediately to mind when discussing a breeding management program based solely upon behavior. What about the bitch that never becomes receptive, or will only stand for a day or just a few hours? When faced with a problem bitch of this nature, you will have to rely on other signs and information. Vaginal cytology is the best because you will not be fooled about the bitch's

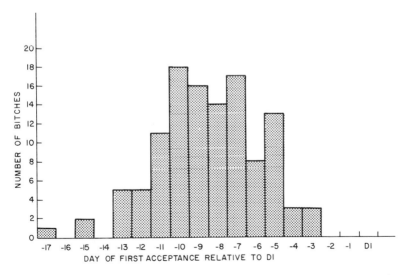

**Fig. 10.2 Onset of Estrus (First Acceptance) Compared With D1** *in a study of 116 Beagles. On average acceptance occured 8.5 days before diestrus, which is 2.5 days before ovulation. A fairly wide range was observed, however.*

*Reprinted with permission from Am.J.Vet.Res. Vol 35, No 3, 1974.*

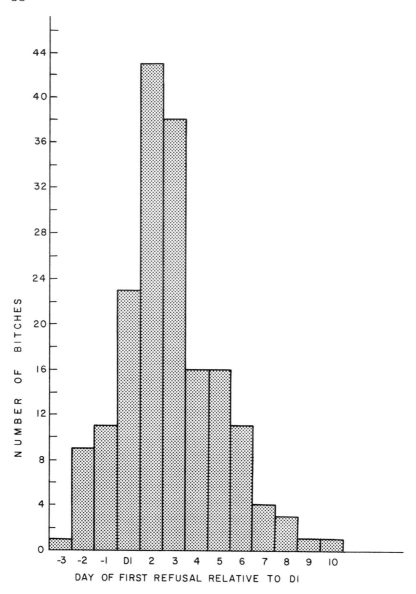

**Fig. 10.3 Day of First Refusal Compared With D1** *in a group of 177 Beagles. On average first refusal occurred 2.7 days into diestrus. If breeding is delayed until late in the cycle, it is likely that an infertile breeding may occur.*

Reprinted with permission from Am.J.Vet.Res, Vol 35, No 3, 1974.

stage of heat, regardless of her behavior. If smears tell you that the time is appropriate to breed, you may need to either force breed the uncooperative bitch or perform an artificial insemination. Other signs, including discharge and appearance of the vulva, must be considered as well.

A potential pitfall exists in the use of behavior to determine breeding, and it is related to the fact that most bitches will continue to breed for several days after ovulation and for an average of three days into diestrus. (Figure 3). If circumstances do not permit you to check when she actually began estrus, you must be aware that it is possible in most bitches to breed too late. If a first breeding is done fairly late in estrus, the chance of a miss is very real, and a vaginal smear should always be done to be sure that the vaginal lining is still cornified and that the breeding will not be wasted. Don't assume that because the bitch will still stand, she will conceive.

# Vaginal Cytology

The use of vaginal smears has become indispensable to me in breeding management. No matter what other signs are present and no matter what the bitch's behavior, you can be absolutely confident of your breeding if the vaginal smear says "breed." When a full series of smears is made, the simplest program would involve waiting until the smear has been fully cornified for four days before attempting to breed. When shipping is involved, it is best to wait until the smear has been cornified for a day or two, then ship the bitch. This will allow enough time for her to recover from her trip before breeding, yet will not get her to the stud too early. When the bitch becomes receptive, breed immediately, then do a second breeding two, three, or four days later, depending on how long the smear has been cornified at the time. A more detailed discussion of this principle may help.

Successful breeding has been shown to occur in bitches bred from four days before until three days following ovulation. Since fertilization occurs three days after ovulation in the bitch, this means that maximum conception rates can be achieved by breeding anytime during the seven days prior to maturation of the oocytes. Expressed another way, the bitch's fertile period extends from D-10 to D-3, ten days to three days before the onset of diestrus. The vaginal smear will be of the fully mature, cornified pattern during this entire fertile period in normal bitches.

An important principle of management, then, is that breeding must occur while the smear is fully cornified, regardless of other observations, including behavior. You need not pinpoint ovulation, and in fact smears cannot be used to determine the time of ovulation except in retrospect, after D1 has been determined. In a bitch whose smear remains fully cornified for ten days, the desired breeding time (equivalent to D-10 until D-3) will be the first to the eighth day of the cornified smear (Figure 4). In the bitch with an eleven-day cornified period, the corresponding fertile period will be the second to ninth day of her cornified smear. A similar estimate can be made for bitches whose smears remain cornified for twelve, thirteen, and fourteen days. Any normal bitch, whether her period of cornification lasts ten, eleven, twelve, thirteen or fourteen days, can be bred between the fifth and eighth days of complete cornification with excellent prospects for conception. This information is useful in a prospective way and can be helpful in management when other signs during the estrous cycle are misleading or difficult to interpret. Each bitch tends to have a constant duration of cornification from one cycle to the next.

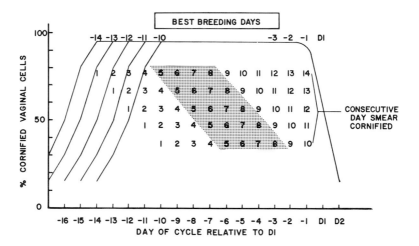

**Fig. 10.4 Use of Vaginal Cytology to Determine Breeding Time.** *First, daily smears are used to determine the beginning of complete vaginal cornification. Bitches with a period of cornification lasting 10 to 14 days have a best breeding time which can be determined from the chart. Any normal bitch should become pregnant if bred on her 5th to 8th days of cornification (shaded area).*

# Counting the Days of Heat

An old, time-honored method of breeding management is to breed on a given day or days during the heat cycle. Usually it is the tenth or tenth to thirteenth day of heat that is used. When the bitch is normal and average, this plan is highly successful. Some breeders have used only this method for years. From time to time, I hear from a frustrated or bewildered breeder who laments, "I always bred Suzy on her tenth and twelfth day before, but this time she wouldn't stand until the fourteenth day. What's wrong?" Probably nothing is wrong. Suzy is undoubtedly normal, but this time she has deviated a little from average. The actual numerical day of a bitch's heat is never my primary concern when dealing with a breeding problem. So often the owners have missed the first few days anyway. Along with knowing the day of heat, you need to know the appearance of the vulva and the nature of the discharge, whether the bitch is receptive, and what the vaginal smear says. Do not spend a lot of time looking at your calendar; look at your bitch! She knows exactly what she is doing.

# Checking for Color Changes

A client came in to have a bitch checked that had failed to become pregnant on two breedings. This breeder watched for a change in her vaginal discharge from red to amber, waited four days, then bred her. I do not know where this program originated, but suffice it to say that it did not work, and color is a highly unreliable sign. Do not count on it. Look at every aspect of the bitch's heat and not just one feature. Review the chapter on physical aspects of the estrous cycle for further details.

# To Achieve Optimum Fertility

Conception rate refers to the percentage of bitches in a group that conceive a litter, and a single pup constitutes a litter. There is another way of expressing fertility, and it is much harder to achieve because it involves examination of the ovaries, counting CLs, counting the embryos or pups conceived, and then calculating a percentage of eggs that were successfully fertilized. This was done in one group of Beagles with some fascinating results. The bitches bred before ovulation and on the day of ovulation had a fertility rate of 83 to 85 percent of eggs fertilized. The bitches bred one day after ovulation had a 90 percent fertility score, while those bred two days after ovulation had a 96 percent rate (Table 1).

We know that the egg requires approximately three days to mature following ovulation before it can be fertilized. We also know that the sperm survive in the reproductive tract until the eggs are ready to be fertilized. So when a bitch is bred two days following ovulation, the sperm are relatively newer or fresher at the time of fertilization than if she were bred one, two, or more days earlier. It is only conjecture that the freshness of the sperm available increases or optimizes the percentage of successful fertilization. The increased fertility might also be due to the environment in the bitch's tract during the days following ovulation. Whatever the reason, there is a higher fertility rate in bitches bred one and two days after ovulation, and they produce a slightly higher average litter size.

Now we need to examine just what good this information may do us in breeding management. If we could know exactly the day of ovulation, we could wait forty-eight hours to breed and possibly have a little higher average litter size. Unfortunately, it is not that easy to know when ovulation occurs. Hormone assays might tell us and even that is uncertain, but they are not available on an overnight basis. Vaginal cytology can tell us quite accurately in retrospect (ovulation occurs six days before the onset of diestrus), but we can only make

**Table 1. Fertility Estimates in a Study of 72 Beagles.**
*The numbers of corpora lutea ( = eggs ovulated) is compared with the number of live pups to achieve a percentage of eggs successfully fertilized and developed. Optimum fertility was seen when breeding one and two days following ovulation (D-5 and D-4).*

| Day of breeding | No. | Total CL | Total living pups | Pups per CL | Pups per litter |
|---|---|---|---|---|---|
| D-11* | 3 | 20 | 10 | 0.50 | 3.3 |
| D-10 | 10 | 74 | 61 | 0.82 | 6.1 |
| D-9 | 8 | 53 | 46 | 0.87 | 5.8 |
| D-8 | 7 | 43 | 37 | 0.86 | 5.3 |
| D-7 | 13 | 92 | 77 | 0.84 | 5.9 |
| D-6 | 13 | 84 | 69 | 0.82 | 5.3 |
| D-5 | 6 | 42 | 38 | 0.90 | 6.3 |
| D-4 | 8 | 57 | 55 | 0.96 | 6.9 |
| D-3 | 4 | 32 | 20 | 0.63 | 5.0 |

*D refers to time relative to onset of diestrus.

Reprinted with permission from Am.J.Vet.Res. Vol 35, No 3, 1974.

a guess during heat because we cannot be sure how long a bitch's smear will remain cornified—the normal range being ten to fourteen days. So the only thing to do in order to achieve a breeding on D-4 or D-5 is breed daily or every other day during the middle to latter part of estrus. At least that is all we can do until someone develops a quick, reliable test for detecting ovulation.

# *Tes-Tape®

The use of glucose-sensitive strips for detecting glucose in vaginal secretions has been advocated by some for determining breeding time. The strips are designed to detect glucose in urine and are used by diabetics to help monitor their glucose levels and insulin requirements. The test strip is a thin strip of paper that turns from yellow to green on contact with fluid containing glucose. When a piece of the paper is placed in the vagina, it will turn green if glucose is present. I tested this procedure on several Beagle and Sheltie bitches and found that the results were extremely variable and unpredictable. There was little correlation between the day the strip turned green and what was later determined to be the day of ovulation. I have seen this method published on numerous occasions in lay journals, dog publications, and kennel club newsletters, but never in professional veterinary or research publications. I do not recommend the method and do not believe that it will help your breeding management, except perhaps by coincidence when the tape turns green on a day that is a good time to breed. Table 2 shows the results of *Tes-Tape® in three Sheltie bitches

**Table 2. Results of *Tes-Tape® Testing In Three Sheltie Bitches.**

| Day From D1 | -17 | -16 | -15 | -14 | -13 | -12 | -11 | -10 | -9 | -8 | -7 | -6 | -5 | -4 | -3 | -2 | -1 | D1 | D2 | D3 |
|---|---|---|---|---|---|---|---|---|---|---|---|---|---|---|---|---|---|---|---|---|
| **Melanie** Day of heat: | | | | | | | | 6 | 7 | 8 | 9 | 10 | 11 | 12 | 13 | 14 | 15 | 16 | 17 | 18 |
| Tes-Tape result: | | | | | | | | - | - | - | - | - | + | ++ | - | - | - | ** | ** | |
| **Dilly** | 1 | 2 | 3 | 4 | 5 | 6 | 7 | 8 | 9 | 10 | 11 | 12 | 13 | 14 | 15 | 16 | 17 | 18 | 19 | 20 |
| | - | - | - | - | + | - | - | - | - | ++ | - | - | - | + | - | ++ | + | + | - | + |
| **Laurel** | | | | | | 5 | 6 | 7 | 8 | 9 | 10 | 11 | 12 | 13 | 14 | 15 | 16 | 17 | 18 | 19 |
| | | | | | | ++ | ++ | +++ | # | ++ | ++ | - | ++ | ++ | + | + | + | + | + | + |

*Tes-Tape®  Eli Lilly and Company, Indianapolis, IN 46285.
Results: (-) no change in tape color, ( + ) barely detectable green color or green along one edge of tape, ( ++ ) light green color, ( # ) dark green color.
** No vaginal discharge to test.

# Influencing Sex Ratio

People have always been fascinated by the idea that there may be some way to change or manipulate the sex ratio. In dogs, approximately 55 percent of the pups whelped are male and 45 percent are female. As in almost all animals, the sex is determined by the sperm that fertilizes each ovum. Females are genetically XX in their sex chromosome makeup, and males are XY. Each ovum thus carries the X chromosome, and each sperm carries either X or Y. The X sperm will produce XX, a female, and the Y sperm will produce XY, a male. Why the normal sex ratio is 55 percent males in dogs is really not understood. It seems that breeders are always longing for more female puppies. To my knowledge, there has never been a breeding method proven to really influence the sex ratio. I would like to present the results of one study that I did which served to tantalize my imagination, but which unfortunately did not involve enough bitches to give

**Table 3. Sex Ratio at Birth.**
*Each Bitch was bred only once at a known day relative to the onset of diestrus.*

| Breeding Day | Number of Litters | Total Males | Total Females | % of Females | % of Litters with more Females than Males |
|---|---|---|---|---|---|
| D-11,-12 | 6 | 18 | 16 | 47% | 33% |
| D-10 | 11 | 36 | 26 | 42% | 18% |
| D-9 | 6 | 18 | 21 | 54% | 67% |
| D-8 | 10 | 33 | 29 | 47% | 40% |
| D-7 | 17 | 61 | 45 | 42% | 35% |
| D-6 | 9 | 31 | 26 | 46% | 33% |
| D-5 | 9 | 24 | 40 | 63% | 67% |
| D-4 | 6 | 12 | 23 | 66% | 100% |

significant numbers. The bitches bred at and before ovulation had 55 percent male pups, the expected ratio for Beagles. The bitches bred one to two days following ovulation had 35 percent male pups, a shift in the direction of females (Table 3). I do not know if what I saw was true, or only a coincidence, but would like to present it to tantalize your imaginations as it did mine! The reason for a higher percentage of females when breeding took place on D-9 is not clear and may mean that all of these figures are merely coincidental. I have seldom since had information on a bitch that could be added to this list. Most were either bred more than once, or not enough smears were available to determine ovulation, or both, and unless the same criteria are observed (i.e., a single breeding at a known time relative to ovulation), the results are not comparable.

# Litter Size Related to Age

It would be considered normal for a bitch to have a slightly smaller litter the first time than in the following litters. The peak of reproductive efficiency occurs between two and six years of age. After six years, litter size begins to decline, and bitches older than nine are expected to have much smaller litters than during their prime. The reason for this might be because of uterine changes, such as scar formation at the area of previous placental attachments, or chronic low-grade endometritis. I am often asked how old can a bitch be and still be bred. The answer depends on the breed and the general health and fitness of the individual. Many individuals could have a healthy, although perhaps small, litter at the age of ten. AKC requires special certification before registering a litter from a bitch more than twelve years of age.

# Mismatings

Heaven forbid that it should ever happen to your bitch but some of us have had the misfortune of having one of our bitches accidentally bred by a stray that jumped the fence or by a dog of a different breed. What do you do when this happens?

There are three options, perhaps four, open to us. The first, although not acceptable to AKC, would be to immediately mate the bitch to a male of her breed and hope that part or all of the litter would be purebred. If so, at least something might be salvaged from the misfortune. A second option would be to have the bitch spayed, thus terminating the pregnancy. In the case of a bitch that you intended to retire or not breed anyway, this would be a desirable solution. Third, you could consider an aborting injection of estrogens immediately after the mismating. There are some potential dangers involved in this that must be considered in the case of a valuable breeding animal. The last option would be to allow the bitch to carry and whelp the litter. This would be a safe route, as far as the bitch's health is concerned, but leads to yet another decision. The pups could be reared, and hopefully they would find suitable homes. That is not easy for mongrels, as they are overpopulated everywhere. Alternatively, the pups could be humanely destroyed immediately after whelping. This solution should not be overlooked; it may be more kind than any other in some cases.

What about the future breeding of your mismated bitch? The answer to this should be clear to you by now, but let me set it down so as to eliminate any chance of misinterpretation. Your bitch's future heats, future matings, future pregnancies, and future pups would not be influenced in any way by a mix-breed litter. Every ovulation is unique, every pup conceived is unique, and every pregnancy is totally independent of what went before.

Chapter Eleven

# A LITTLE HELP NEVER HURT: ARTIFICIAL INSEMINATION

There are times when a natural mating is not possible or desirable for various reasons, and artificial insemination (AI) can be an extremely valuable tool. Some of the indications would be a physical disability, such as an injury, to either the male or female, lack of libido on the part of the male, lack of interest in the particular bitch, immaturity, inexperience, or insufficient aggressiveness on the part of the male. Or it could be fear of spreading infection, aggressiveness on the part of the female, lack of interest in the particular male, or insufficient size of the vulva or stricture that would make mating impossible. Behavioral problems could also make the bitch unwilling to breed.

The procedure is quite simple. First, the bitch should be checked thoroughly to be sure that she is in the proper stage of her cycle to be bred. A vaginal smear should be considered mandatory. Remember—AI is being done because some physical or behavioral aspect of the cycle demands it, and it would be foolish to proceed without knowing that you have a reasonable chance of success.

To collect semen from the male, it is best to have the bitch present to act as a teaser. The same is true when collecting semen from a male for semen evaluation. The male should be allowed to sniff the bitch, lick her, and even mount her if possible. These activities stimulate the male and facilitate collection. The person collecting the male should take time to get him used to being touched and handled. The collector must talk to the dog and try to get the dog to relax while being petted, and then while his abdomen and prepuce are being touched.

Semen collection is achieved by physical stimulation and manipulation of the penis. First, the body and bulb of the penis can be gently massaged through the prepuce. As soon as the bulb begins to enlarge, the sheath is slipped back behind the bulb and the penis is grasped with the thumb and first finger encircling the bulb. In most dogs, simple, continuous pressure behind the bulb will stimulate ejaculation, but in others some intermittent pressure by gently squeezing behind the bulb is helpful. In still other individuals, some massage along the body of the penis helps. Many dogs will have some involuntary thrusting activity during ejaculation, but some will not. It does not appear to be necessary for ejaculation.

Ejaculation begins with delivery of the first seminal fraction while the bulb and the body of the penis are enlarging and achieving erection. The first fraction of the semen is relatively sperm-free and should appear clear and colorless. The second or sperm-rich fraction follows and is cloudy white. Following is the third fraction, which is again colorless and relatively sperm-free. It is not necessary to collect all of the third fraction, only enough to be sure that all of the sperm-rich fraction has been collected. When collection is complete, the dog's penis can simply be released from the hand and the dog allowed to stand until the erection has decreased. The volume of semen varies with the size of the dog and between individuals of the same breed. Usually one to ten milliliters can be collected.

It may take some time for the penis to decrease to normal size, and it is important to be sure that the dog is quiet during this time so that he won't injure himself. Ejaculation of the third fraction will continue intermittently. After the penis has returned to the sheath, it is important to check to be sure that is has fully retracted. In some cases the skin at the tip of the prepuce rolls inward as the penis retracts and can cause some discomfort or exposure of the tip of the penis.

Collection equipment for artificial insemination can be very simple or extremely sophisticated. Generally, you can use whatever is available as long as some important principles are followed. The container for collection must be scrupulously clean, preferably sterilized to avoid the possibility of spreading infection, and absolutely dry and slightly warm. A small drop of water unnoticed in the bottom of the collecting vessel will kill the sperm instantly because of disturbance of the osmotic balance. This is not true of physiologic saline solution (0.9 percent NaCl), such as is used to make vaginal smears, because it is balanced to the body and sperm's osmolarity (concentration of solids in fluid). In fact, 0.9 percent NaCl can be used to dilute and extend semen. The equipment should be warmed to body temperature or nearly so to avoid temperature shock to the sperm. In actual practice, holding the container in your hand seems to be sufficient.

Semen can be collected into an open-top container, such as a glass, cup, or test tube, or it can be collected into a test tube attached to a soft rubber, funnel-shaped sheath. The rubber sheath surrounds the penis during collection and prevents any loss of the collected sam-

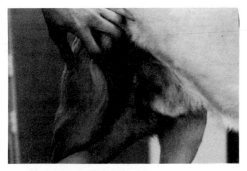

**Fig. 11.1 Collection Begins With Massage of the Penis Through the Prepuce.**

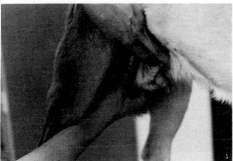

**Fig. 11.2 As soon as the Bulb Begins to Enlarge,** *the prepuce is slipped back so that the bulb is exposed.*

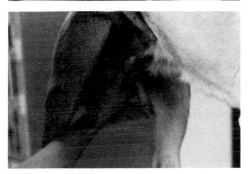

**Fig. 11.3 Constant or Intermittent Pressure** *of the hand encircling the penis behind the bulb should stimulate ejaculation.*

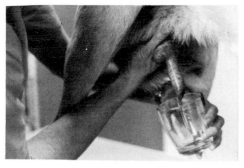

**Fig. 11.4 The First Seminal Fraction** *is ejaculated while the penis is becoming erect.*

**Fig. 11.5 Involuntary Pelvic Thrusting** *often occurs in the later stages of erection. Care must be taken at this point to avoid injury to the penis.*

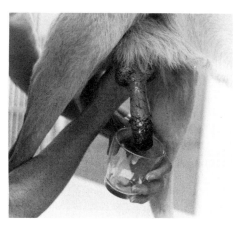

**Fig. 11.6 The Sperm-Rich Second Fraction** *is ejaculated during and immediately following pelvic thrusting. The penis is fully engorged at this time.*

**Fig. 11.7 A Small Amount of Clear Third Fraction** *can be collected to ensure that all of the second fraction has been collected. The third fraction is not needed for fertility. Some dogs may seem more comfortable if the penis is held back in the natural position it would have during a coital tie.*

ple. The sample then is collected directly into the test tube at the bottom of the funnel-shaped device.

Immediately after collection, the semen should be examined for concentration, motility, and abnormalities. Just as it is imperative to examine the bitch to determine that the effort to do the AI is justifiable, it is equally important to be sure that the dog's semen is of good quality and worth the effort to inseminate.

Insemination is done using a sterile, dry syringe and a plastic inseminating tube six to eight inches in length with an adapter on one end to attach to the syringe. The tubes can be obtained from veterinary or dairy supply companies and are designed primarily for use in AI in cattle.

The semen is drawn slowly and gently up the inseminating pipette into the syringe. Then it should be immediately inseminated. A small amount of surgical or obstetrical lubricant is applied to the tip of the tube. The bitch to be inseminated should be held with her hindquarters elevated, preferably with an assistant seated in a chair. The bitch's front feet should be on the ground and her rear feet on the seated person's knee, or draped over the knee or leg. The position will have to be held for at least five minutes, so it should be comfortable for all concerned.

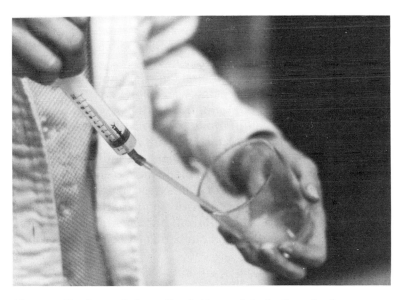

Fig. 11.8 The Semen is Immediately Drawn Into the Inseminating Pipette. *Insemination is done as soon as a drop of semen has been evaluated and found to be of good quality.*

The inseminating pipette is gently inserted into the bitch's vagina as far as it can be passed. Keep in mind the arch over the pelvis while inserting the pipette. The pipette can be inserted four to five inches in small bitches and the full six- to eight-inch length in larger individuals. The semen is then slowly and gently injected. The site for insemination is the cranial portion of the vagina, near the cervix. Because of the anatomy of the cervix, it is impossible to pass the pipette into the uterus, which probably would be the ideal site for insemination were it possible.

Ater the semen has been deposited, the bitch should be held with her hindquarters elevated for at least five minutes. This may not be needed, but it presumably aids in movement of semen into the uterus via gravity. I also like to put on a sterile glove, lubricate one finger, and insert the finger into the caudal vagina for about five minutes. The pressure of the finger simulates the natural pressure during the coital tie. Most of the time it also stimulates some contractions in the muscular wall of the vagina and hopefully helps move the semen into the uterus via peristaltic contractions of the organs. After the five minutes, the bitch can be let down to resume normal activity. Whether or not she is confined or allowed to exercise or urinate should have no effect upon the outcome of the artificial insemination.

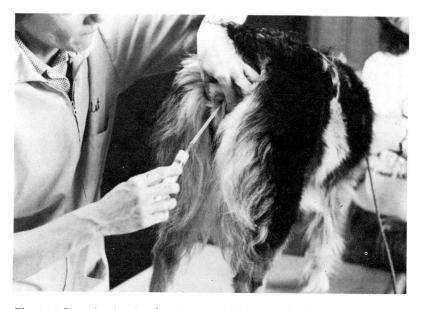

**Fig. 11.9 Insemination Involves Inserting the Pipette Gently Into the Vagina.** *The tip is lightly lubricated with obstetrical lubricant (water soluble, non-spermacidal), The bitch's smear has been checked for stage of cycle and signs of infection.*

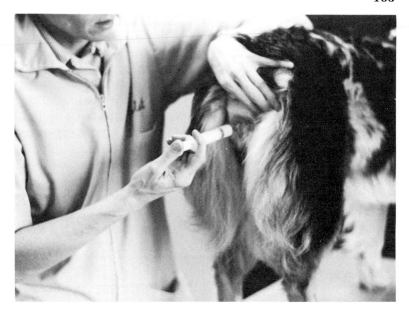

**Fig. 11.10 The Semen is Slowly Injected Into the** *cranial vagina. The pipette was inserted approximately 5 inches in this bitch.*

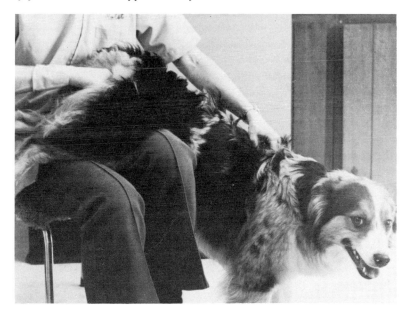

**Fig. 11.11 After Insemination the Bitch is Held for at Least Five Minutes** *with her rear quarters elevated. A gloved finger is held in the vagina to simulate the pressure of a natural breeding which stimulates vaginal muscular contractions. Gravity and peristaltic activity help move the semen into the uterus.*

Artificial insemination can have the same rate of success as natural breeding. In order to achieve this rate of success, however, all the principles mentioned previously must be followed. I do not even bother in inseminate, for example, when the dog's semen contains no live sperm or when the bitch is not in the proper stage for breeding. If AI is done in that kind of hopeless situation, of course it will come out looking very bad statistically. Semen can be mixed with nutrient extenders, cooled, and stored for up to two days at refrigerator temperatures and still retain some fertility.

# Frozen Semen

The technology for the freezing and thawing of semen in dogs was reportedly first successful when a bitch at the University of Oregon Medical School whelped a litter in May 1969 that was conceived after insemination with frozen semen stored for six months. Since that time, the techniques have been variously successful, and in March 1981, the American Kennel Club first announced that it would allow registration of litters resulting from the use of frozen semen. In 1982, the first ACK registered litters were produced from frozen semen. Hopefully, in time the techniques will be perfected, but as of this writing, successful conceptions have been few. The use of frozen semen would decrease the need to ship valuable animals long distances for breeding, and it would greatly extend the breeding life span of our best males.

Chapter Twelve

# THE WHY AND WHY NOT OF HORMONES: HORMONE THERAPY

Owners of breeding animals, when faced with reproductive problems in their valuable stock, often ask if there isn't some kind of hormone that can be given by their veterinarian to make things right. The answer is never simple and is usually "No." Hormones are chemicals normally produced in glands. They are released into the bloodstream and have some kind of influence on other cells, tissues, or organs of the body. Review the section on the hormonal control of the estrous cycle to remind yourself how complex normal hormonal relations are and how finely balanced a normally functioning reproductive system is. Any kind of hormonal manipulation should be attempted only by a veterinarian, preferably one with special interest and training in reproductive problems.

## Induction of Estrus

One therapy commonly sought by breeders is to induce heat in a bitch that is either older than twelve months before reaching puberty or that has gone longer than the owner likes between heat cycles. In most cases, the age at puberty and the interestrous interval are normal. Age at first estrus in a few extreme cases can be as late as two and one-half years. Intervals between heat cycles are often twelve to fourteen months. At the present time, there are no reliable and safe methods for inducing a bitch to come into estrus. Various schemes, protocols, and treatment programs have been tried and are still being tested, but none has yet proven its worth and safety. Generally, these protocols involve the use of the gonadotropins, FSH, and LH, but the results have been unpredictable.

# Abortion

Estrogen can be used to terminate pregnancy if given during estrus within a short time after a mismating. Generally it is administered in the form of estradiol cypionate (ECP) by intramuscular injection and always by a veterinarian. Given properly during estrus, it elevates the blood estrogen level in the bitch at a time when it would normally be dropping. The elevation of estrogen at this time causes the functional lock of the ova in the oviduct to be prolonged. The ova are held in the oviduct at a time when they would normally pass into the uterus, and they degenerate, thus causing the pregnancy to be terminated. Prolonged elevation of estrogen also makes the estrous cycle longer; the bitch will stay in heat for one to two weeks longer than she would have otherwise. She could probably be bred again, but because of the tubal lock should not conceive again during the same heat cycle. Therefore, estrogen injection should not be repeated in the event of a second misalliance.

The metabolism of ECP is slow and its effects are prolonged so that the drug is potentially toxic. The toxic effect of an overdose is depression of bone marrow with a potentially fatal aplastic anemia. Estrogen also increases receptors for progesterone in the uterus which will potentiate the effects of progesterone on the uterus. The overstimulation of progesterone can cause cystic changes in the uterus which can lead to pyometra. Such complications are quite uncommon following the proper use of estrogen for mismatings, yet any risk should be taken into consideration. A valuable brood bitch probably should be allowed to whelp the unwanted litter and the pups or most of them disposed of at the time of whelping. It would be a temporary setback to her show or breeding career, but relatively free of risk.

# Estrogen—Other Therapeutic Uses

## Spay Incontinence

Some bitches experience incontinence, or dribbling of urine, usually during sleep, after they are spayed. This problem can usually be controlled with the use of orally administered diethylstilbestrol (DES) in small doses. Toxicity is not likely to occur with DES given orally, and the dosage is usually small and given at weekly intervals.

## Male Dogs

ECP and DES can be used to control certain types of androgen-dependent perianal tumors in male dogs and benign enlargements of the prostate when castration is not feasible. Estrogens decrease production of androgens by a feedback inhibition of the pituitary gland. In some cases a progestin rather than an estrogen may also be used in these dogs. Megestrol acetate, also known as *Ovaban®*, has been used for this purpose and appears to have fewer undesirable side effects than do estrogens.

# Prevention of Estrus

Various progesterone compounds have been used from time to time to prevent a bitch from coming in heat. These compounds act by a negative feedback mechanism to suppress activity of the pituitary gland and prevent a normal estrous cycle. Unfortunately, another natural effect of progestins is stimulation of the uterus, and such drugs can cause endometrial hyperplasia and pyometra. So, although these compounds are effective contraceptives, they are not approved for use in dogs!

Megestrol acetate (*Ovaban®*) is now the only approved progestational compound for use in dogs in the United States. The drug is a short-acting progesterone derivative and acts by inhibiting the gonadotropin (LH, FSH) release from the pituitary gland. It can be given for thirty-two consecutive days during anestrus at a low dosage, or for eight consecutive days during early proestrus at a higher dosage. According to the manufacturer's recommendations, megestrol acetate should not be used before the bitch's first heat cycle or for more than two consecutive heats. When given according to directions, the incidence of pyometra is reportedly no higher than in bitches given no hormonal therapy. Side effects include weight gain, mammary enlargement, and listlessness. While the drug's manufacturers do not mention this feature, I have seen bitches exhibit normal estrous behavior and be extremely attractive to males when given *Ovaban®* in proestrus. It presumably occurs because the bitch is given a progesterone-like compound when her estrogen level is high. This is the hormone combination which causes attraction of males and mating behavior during a normal heat, so it is not surprising that the same occurs with the drug given at the appropriate time. Even though mating may occur, pregnancy would not because ovulation is blocked.

Androgens have been used to prevent estrus in bitches, and they, like progestogens, act to inhibit release of gonadotropins from the pituitary gland. Androgens are also anabolic (building up of body tissues) and are used in some racing dogs to help improve their performance. They are sometimes used by human athletes for the same reason.

Androgens can be given orally, usually as testosterone. Experimentally, implants of testosterone have been used to inhibit fertility in bitches. Side effects include enlargement of the clitoris, once in a while with formation of an os clitoris, which is analogous to the male's os penis. The enlarged clitoris can act as a source of irritation and lead to vaginitis. If an os clitoris has developed, it may have to be removed surgically to relieve the problem. Urogenital abnormalities may occur in female pups born to bitches given testosterone, and some adverse effects on liver functions have been observed. The use of testosterone for contraception is not approved in the United States, yet it has been used for this purpose in racing Greyhounds for many years.

Mibolerone (*Cheque®*) is the only androgenic compound approved for fertility control in dogs in the United States. It is not approved, however, for use in breeding animals. Reduced fertility has been observed in animals that have been treated with mibolerone. It is given daily for up to twenty-four months and is in liquid form. Side effects are the same as those seen with testosterone: clitoral enlargement, secondary vaginitis, urogenital abnormalities in female fetuses, and elevation of liver enzymes. It has also been reported to interfere with milk production so should not be given to lactating bitches. It has been used to decrease lactation in some cases of pseudopregnancy. Mibolerone should not be given to bitches before their first heat, because androgens can cause premature closure of the growth plates of the long bones in the pup's body. Bitches begin to cycle at a variable time after withdrawal of the drug, from a few days to several months.

Perhaps you have been impressed with the number of undesirable side effects of the hormones used to prevent estrus in the bitch. Yet, these drugs are effective and hopefully have been helpful in accomplishing their intended purpose: to alleviate the problem of the mass production of unplanned and unwanted litters. To a serious breeder of valuable purebred stock, it is doubtful that the benefit of keeping a bitch out of heat at an inconvenient time is worth the risk. Hormonal manipulation of the estrous cycle is generally not advised in animals that are considered valuable as breeding stock.

# To Improve Performance or Litter Size

Some breeders have tried to improve their bitch's conception rate and/or litter size by giving injections of the pituitary gonadotropins.

The hormone preparations are made from pituitary glands or blood of domestic animals such as sheep, cattle, or horses. Purified FSH is available. Human chorionic gonadotropin has also been used. Pregnant mare's serum has been used, and it is mainly FSH with a little LH activity. Theoretically, given at the appropriate time, these hormones could stimulate the ovaries to produce more follicles, ovulate more ova, and result in a larger litter size. FSH is normally low during proestrus (see Chapter 5), so the actual effectiveness of such treatment is uncertain. A preparation with LH-like activity is human chorionic gonadotropin. It has been used to stimulate ovulation, such as in the case of cystic follicles.

A major drawback to the use of such hormones is that they are proteins, or the preparations contain proteins that are foreign to the bitch's own body. Her immune system will produce antibodies against the injected material, and these can result in disastrous consequences when the treatment is repeated at a later time. A second injection in a sensitized individual can cause a severe, even fatal, immune rejection response. In the case of LH preparations from sheep or cattle, antibodies against the injected LH may cross-react with the bitch's own natural LH and cause sterility. Unfortunately, canine-origin LH is not available commercially.

Considering the uncertainty of benefit and the unpredictability of results, and the chance of undesirable harmful effects, hormone treatment to improve fertility should be avoided in most valuable breeding animals.

# Treatment of Short-Cycling

Bitches that cycle every three to four months often, although not always, have fertility problems. A three- to four-month cycle probably is not normal, but the cause of the condition is unknown. What to do about it is also a frustration. In a normal estrous cycle, it is believed that the uterus requires up to 150 days to complete its involution and to be prepared for the next cycle and pregnancy. Allowing for proestrus and estrus, a minimum "normal" estrous cycle length would be five to five and one-half months. Based upon the presumption that the short-cycling bitch fails to produce a litter because her uterus does not have adequate time to regenerate between ovulations, mibolerone (Cheque®) has been used to keep bitches from coming in heat. It is administered daily until the bitch has been kept out of heat for at least six months. This use of mibolerone has not been tested in a disciplined, controlled study, but in some cases has been helpful.

# Treatment of Male Infertility

The pituitary gonadotropins, FSH and LH, have been used to treat azoospermia (no spermatozoa) and oligospermia (few spermatozoa) in males. Theoretically, these hormones might be beneficial in stimulating the seminiferous tubules and interstitial cells to activity. Unfortunately, the treatments are usually of no benefit. At Colorado State University during a three-year period, 90 percent of the males tested because of azoospermia had elevated LH and/or FSH already, which suggests that the problem is caused by something besides a lack of pituitary gonadotropins.

Other hormones have been used to treat male infertility, but usually without much success. Thyroid hormone has been used to stimulate general metabolism. It has also been used with corticosteroids, on the assumption that the problem involves inflammation or immune-mediated damage to the testes. Even diethylstilbestrol has been used in some therapeutic protocols. Generally, azoospermia and atrophy of the testes have resisted treatment. Testosterone is not only unnecessary, but it is contraindicated in such cases because of the negative feedback mechanism that decreases secretion of gonadotropins from the pituitary. The testicular problem might actually be aggravated.

# Treatment of Pyometra

Until quite recently, virtually the only treatment in cases of pyometra was hysterectomy. Prostaglandins have been used with some success in cases of open-cervix pyometra. Prostaglandins are chemicals present in many tissues and have a wide variety of effects throughout the body. Prostaglandin-F2 $\propto$ causes regression of the corpus luteum in several species of domestic animals. Under the influence of this prostaglandin, the progesterone level in the blood falls. Pyometra occurs when progesterone levels are high, and prostagland-F2 $\propto$ has proved to be helpful in treating an otherwise hopelessly diseased uterus. Several injections are given at twelve hour intervals. When the treatment is successful, the uterus is emptied and the corpora lutea regress.

Prostaglandin-F2 $\propto$ has also been used experimentally to cause abortion in bitches but is not currently approved for this use.

# Prevention of Abortion

Progesterone from the corpora lutea is needed to maintain pregnancy in the bitch until the last day before whelping. If the bitch produces inadequate progesterone or the corpora lutea regress prematurely, the pregnancy will fail, and the conceptuses will be reabsorbed or aborted, depending upon their stage of development. In some cases, supplemental progesterone has been beneficial in allowing these bitches to carry a litter to term. The repositol type of progesterone is used and is injected at intervals throughout the last half of pregnancy. Of course, in cases of abortion, other causes including infections and brucellosis must be ruled out by appropriate testing. Progesterone declines prior to any abortion, and progesterone supplementation in the face of infection would be contraindicated. It could aggravate the infection.

Chapter Thirteen

# FEEDING TIPS FOR HEALTHY PUPS: NUTRITION

Many excellent and lengthy discourses have been written on canine nutrition and on the various nutrients and requirements. This chapter will summarize the guidelines for feeding and management during breeding, pregnancy, lactation, and early days of puppy development.

Good nutrition is essential to the production of healthy puppies. Before breeding, a bitch should be in excellent health. She should be neither too fat nor too thin and be free from parasites, skin problems, and any other diseases.

Many fine dog foods, and many fad diets, supplements, and special foods are available. The best feeding program is the one that keeps your dog in good condition and that is based upon a top-quality dog food with minimal supplementation. The dog food industry is highly competitive; the major manufacturers have invested millions in testing and quality control. You should stick with a dog food that is made by a company with a good research program and feeding trials. Avoid feeds from companies that cannot provide information on testing and feeding trials. The best foods are ones that list meat or animal protein as one of the first four ingredients on the package label.

Supplements should be kept to a minimum. If your dog cannot maintain condition without extensive supplementation, you really ought to consider changing to a better-quality feed. Supplements should never exceed 25 percent of the dog's daily feed, and preferably not more than 10 percent. Animal protein, including meat, milk, cheese, eggs, and liver, are good supplements, especially during pregnancy and lactation. Meats and liver are more nutritious when fed raw, and eggs can be fed either cooked or raw. Raw egg whites contain an enzyme that destroys biotin, an essential vitamin, but this is not a problem unless a lot of eggs are fed. Supplements of single vitamins are seldom needed, because commercial feeds contain adequate amounts and in balanced proportions. Multiple vitamin and mineral supplements are useful at times and are almost always preferred over single vitamins or small groups of vitamins. Calcium and phosphorus supplementation is seldom indicated. Oversupplementation of these minerals is potentially harmful during pregnancy and lactation because it may predispose the bitch to eclampsia. Growing pups almost never need additional calcium and phosphorus, even the largest breeds. Their normal diet contains adequate amounts in a balanced ratio, if it is a quality diet.

# Feeding and Care During Pregnancy

A pregnant bitch should be fed a good-quality feed containing at least 25 percent protein on a dry matter basis, preferably 28 to even 30 percent for toy breeds. At least 10 percent of her calories should be from good-quality animal protein. During the first four to five weeks of gestation, the amount fed should be the same as her normal maintenance diet. The pups do not make significant demands on her at this time, and being overfed or fed to gain weight can be detrimental to her pregnancy. The rate of fetal reabsorption is believed to be higher in bitches that are fed excessively in the few weeks after breeding. Her normal feeding schedule and routine should be maintained.

After five weeks postbreeding, it will be necessary to gradually increase the amount fed, and this in proportion to the bitch's body weight gain. She will gain more and gain faster with a larger litter, and it should be obvious around five to six weeks whether her litter is small, medium, or large. Do not feed to push for tremendous weight gains, even if the bitch is carrying a large litter. By the time she is near term, she should be getting 10 to 50 percent more than her normal maintenance ration.

Supplementation during pregnancy is a debatable topic, but most commercial feeds can be helped by adding some animal protein. If the bitch is on a diet especially formulated for pregnancy and lactation, she probably doesn't need supplements of any kind. If she is on a maintenance type of diet, some supplementation is in order. A balanced multiple vitamin–mineral supplement can be given, starting at four to five weeks after breeding, once a day, and continuing until at least a week after the litter is weaned. An animal-source protein supplement is also safe and helpful. The birth of strong and vigorous pups has been shown to depend on adequate animal protein. And puppies that have had adequate amounts have fewer parasite problems. Some source of high-quality animal protein can be given such as raw liver, eggs, cottage cheese, lean meat, or fish meal. A small amount can be given every other day from four to six weeks gestation, and a daily supplement from six weeks until weaning. I have had excellent performance in my own bitches using liver and eggs plus a vitamin supplement, alternating the liver and eggs on a day-to-day basis. Give one ounce liver or one egg to a medium-size dog. Remember that these foods are supplements only and should never exceed 25 percent of the bitch's diet.

Calcium-phosphorus supplements are seldom needed because most dog foods contain enough for reproduction. The extra minerals needed during pregnancy and lactation are provided by the regular diet because it is being fed in greater amounts proportional to the bitch's needs. Excess calcium and phosphorus may predispose the bitch to eclampsia after whelping. When lactation begins, the bitch's body needs to mobilize calcium for milk production. The parathyroid gland is primarily responsible for the mobilization from body-stored calcium, such as from the bones. If the bitch has been given large amounts of calcium during pregnancy, the parathyroid gland, lacking stimulation, becomes relatively inactive. When it is needed later to aid in milk production, the calcium mobilization is inadequate and low calcium results, causing eclampsia. A dog's ability to conserve and utilize calcium depends on its diet. Low-calcium diets will result in more efficient utilization, while with high calcium diets, much will be wasted.

Zinc is important in reproduction. Animals on zinc-deficient diets have been shown to whelp smaller than normal puppies, weak puppies, or puppies that die after birth. Bitches on a low zinc diet also have a higher incidence of uterine inertia, and one might see the final one or two pups in a litter retained under such circumstances. Excessive supplementation of calcium can lead to the same syndrome, because the transport system for zinc may be hindered by excess calcium. Toxic milk and prolonged discharge after whelping due to poor involution of the uterus or placental sites have also been linked to zinc deficiency. Eggs and liver are good sources of supplemental zinc.

Vitamin E deficiency in newborn puppies may result in weak muscles, inability to get up on the legs and walk, and bloating. Chronic infections may be a problem because vitamin E, along with vitamin B6, pantothenic acid, pyridoxine, and choline, is important to the normal functioning of the immune system. Most dog foods are not deficient, but because of the possibility of transport and storage problems in the pregnant bitch, it may be advisable to give some extra vitamin E.

During the last two to three weeks of pregnancy, all bitches should get two meals per day. They may find it difficult to eat their entire ration at once because of the enlarged uterus filling the abdomen. Their actual body weight gain should be only 5 to 10 percent over their normal weight during pregnancy. Most of the weight gained should be accounted for by puppies, fluids, and placental tissue.

# Feeding During Lactation

The most common cause of inadequate lactation is poor nutrition. The nutritional demands are tremendous on a bitch nursing a large litter, and it is during lactation that feeding must be increased. Generally one and one-half times her normal ration is required during the first week of lactation, two times normal the following week, and three times the normal ration during the third to sixth weeks. To accomplish this increased feed, a bitch can be put on self-feeding, with food in front of her all the time. My favorite way is to feed what would be a normal day's ration two or three times a day. As a rule of thumb, the bitch's needs are 25 percent of her normal ration for each nursing pup. Thus, a bitch with a smaller litter might only need one and one-half to two times the normal amount, depending upon her litter size. It is extremely important to give the bitch a diet with high protein and fat to supply the added nutrition needed to make quality milk.

Another way of calculating the bitch's needs would be in terms of extra calories needed. The extra calories needed for milk production are equal to 100 kcal per pound of body weight of the puppies per day. The peak of demand would be at three to three and one-half weeks of age. At that time, the puppies should be fed dog food so that while their weight is increasing, the needs of the bitch would reach a plateau and remain almost constant until weaning. The extra feed required by the pups after three weeks should come from solid food rather than from extra bitch's milk.

Some bitches reportedly experience a change in temperament during the second to third week of lactation. Typically the bitch who was previously calm and relaxed will become antagonistic, disgruntled, or aggressive when her puppies are approached. It has been reported that giving 250 milligrams of vitamin C twice each day for a medium-size bitch should revert her to her normal quiet attitude in a couple of days. Even though the dog does manufacture its own vitamin C, the quantities may not be enough in periods of stress, such as during lactation.

In cases where puppies under three weeks of age are fussy and agitated and don't ever seem to be satisfied and sleep the way they should, vitamin B complex, especially thiamine, may be beneficial. Increased amounts should be given to the bitch. The reason for this beneficial effect is not known.

During the entire period of lactation, vitamin-mineral supplementation should continue, as well as additional protein supplements in the form of liver, eggs, cottage cheese, or meat. If a bitch is fed enough of a good diet, there is no reason for her to lose body weight during lactation. Even with the largest litters, good management can insure a healthy, well-fed litter plus a healthy, well-nourished mother.

# Feeding Puppies

It is important to start introducing the puppies to regular feed at three weeks of age. Between three and six weeks, the pups gradually become accustomed to eating dog food, making weaning an easy and relatively stress-free period for both the bitch and her pups.

I like to start the pups with a shallow pan, such as a glass pie pan, and a liquid bitch's milk replacer. The first time the pups are introduced to the food, it helps to simply place their front feet in the dish. If they are hungry, they naturally begin lapping. After two or three days of liquid replacer, start to add dog food. Use whatever food you plan to feed the puppies later. I don't believe in any kind of transition period using human baby cereals or pablum or any other such food. They are not balanced diets for dogs, and any three-to-four-week-old pup can eat dog food if it is softened to a consistency that can be lapped. By four weeks, the pups should be on softened dog food mixed with milk replacer formula and mashed or blended to roughly the consistency of oatmeal. By five weeks, less liquid is needed, and the pups should be able to pick up and chew softened morsels of feed. By six weeks, they should be eating food that is softened but not necessarily mashed or homogenized. The milk replacer can gradually be decreased so that by six to seven weeks the pups are being fed essentially the same as adults. The pups should be fed three times per day during the whole weaning process from three to six weeks.

Many people feed puppies free choice after weaning, usually as a matter of convenience. I do not recommend this method for two reasons. First, it tends to create a dog that is a nibbler for life and does not readily adjust to eating heartily once or twice a day when meals are presented. If you are ever faced with the need to put weight on a dog for show purposes or whatever, it is very difficult with a dog raised as a nibbler. But a dog that was raised on regularly scheduled meals will usually readily eat what is given him, and if that means a little more than usual to increase weight, he will eat it without question and gain the weight.

The second and more important reason not to feed young puppies free choice is that they generally grow up overweight. It is cute to see fat, roly-poly little babies waddling around the yard, but it is not good for them. Large breeds especially are prone to various bone and joint disorders when fed excessively during their growth period. It has also been shown that excess weight in earlier years shortens the life span. It is much better to control the amount fed and to check the pups daily for body condition, adjust the amount fed as their needs change, and strive for a steady, moderate growth rate. Extensive research and experience have repeatedly proven the folly of pushing the dogs for rapid growth.

# Feeding the Stud Dog

Not much needs to be said about feeding the stud dog. Simply keep him lean and healthy and avoid letting him become overweight.

Chapter Fourteen

# NINE WEEKS TO WAIT: PREGNANCY—THE WHOLE STORY

## Detection and Signs of Pregnancy

Detection of pregnancy in the bitch can be at times frustrating and elusive, and at other times quite simple. Unlike some other animals, most notably humans, no hormonal changes occur that are diagnostic. A pregnant bitch experiences the same levels of progesterone as the one that is not pregnant. No other hormones associated with pregnancy, such as chorionic gonadotropins from the placenta, are present in measurable amounts.

Pregnancy may be quite obvious in its later stages because of enlargement of the abdomen, but we all want to know as early as possible. Knowing earlier helps in management (feeding, etc.), but perhaps more importantly it heightens our anticipation and enjoyment to know that we are indeed going to have our planned litter.

There are actually only two positive signs of pregnancy, plus several secondary or supportive signs. The primary or positive signs are palpation of uterine locules—the enlargements that develop at the site of each implanted embryo—and the movement of fetuses, which can usually be detected during the final ten to fourteen days before whelping. Detection of fetal heartbeats can be included as a subclass of fetal movement. With practice, palpation can be developed into a fine art, as nothing resembles the feel of a pregnant uterus to experienced fingertips. The uterine locules are first recognizable on actual inspection of the uterus about two days after implantation begins, but they are small and cannot be felt externally. When a bitch is spayed at that time, however, it is possible to feel the swellings because they are firm and spherical and slightly larger in diameter compared to the interlocular portions of the uterine horns. The swelling increases daily, and at approximately seven days after implantation, it should be possible to palpate the locules through the body wall. It becomes easier as each day passes, and in another five to six days it should be practical for even the most difficult bitches. All the swellings will be the same size if the pregnancy is normal. Smaller locules are probably nonviable and will reabsorb.

At about the nineteenth day of diestrus (seven days post-implantation, or twenty-six days post-breeding), the locules will be about the size of a small grape. The palpation technique involves placing the thumb and fingers of one hand, or the fingers of opposite hands in large bitches, across the abdomen. The animal should be standing, and you should grasp gently the entire mass of the abdomen. Imagine trying to fill your hand with everything that is in the abdomen. The more relaxed the bitch's abdominal muscles, the easier this will be. Next draw the fingers (or hands) downward, and try to slip the abdominal organs through the fingertips, one after the other. The uterus lies high in the abdomen, above most of the loops of the bowel and the bladder, below the kidneys. When it slips through your fingertips, the pregnant uterus will feel like firm, spherical, smooth lumps. If you can stabilize the horn between the fingertips, you should be able to trace the length of the horn and feel intermittently the locules for more than one pup. I would not recommend trying to count the number of pups, because it is not easy to be sure that you have found them all and have not counted some twice. Also, it may lead to unnecessary trauma to the bitch and actually serves no useful purpose. If you can feel even one, preferably two or three locules, and can feel them on one or two successive attempts, then you have successfully and positively diagnosed pregnancy.

At about four and one-half weeks post-breeding, the locules have grown to a size a little larger than a golf ball. After about five weeks post-breeding, or thirty days into diestrus, the uterine locules lose their tone and shape because of continued enlargement, elongation, and accumulation of fetal fluids. Palpation is much more difficult at five weeks and beyond, and a positive pregnancy diagnosis can often not be made. The softer, more uniformly enlarged uterus feels too much like other organs, such as the intestines.

The ideal time to palpate for pregnancy, then, is approximately twenty-six to thirty-five days post-breeding, or the nineteenth to twenty-eighth day in diestrus.

The second positive sign of pregnancy, fetal movement, can be detected during the last ten to fourteen days of pregnancy if you are patient. The bitch must be relaxed, and sometimes it takes a little time before anything can be detected. Picture in your mind's eye a small pup (fetus) lying within a pouch of fluid, this surrounded by a fairly thick layer of muscle (the uterus), and finally this whole sausage-shaped structure surrounded by more muscle layers and skin (the abdominal wall), which itself is probably thickened by development of the mammary tissue. You are trying to feel movement of the fetus within its little swimming pool through all the surrounding cushioning layers. It becomes easier with each passing day as the fetuses grow larger and stronger. They undoubtedly stretch, kick, and wiggle very much like newborn pups. Detection of fetal heartbeats with the stethoscope may be possible as early as six weeks but requires skill and patience until later, when the fetuses are larger. It is sometimes difficult to detect the heartbeats, but when they can be heard, they sound like a soft, high-pitched, fast ticking.

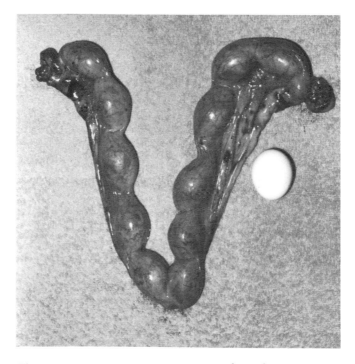

Fig. 14.1 **Pregnant Uterus** *at approximately 28 days post-breeding. Eleven locules are present. A medium-size hen's egg is shown for size comparison. The body of the uterus is not included in this specimen which was removed surgically.*

Ultrasound diagnosis of pregnancy is available at some university veterinary teaching hospitals and is extremely reliable as a diagnostic tool. Pregnancy can be detected as early as uterine locules begin to be defined, and the state of health of the pregnancy, even individual conceptuses, can be determined.

Radiographs (X-rays) can be used to confirm pregnancy but are rarely needed. Simple curiosity about the number of pups is not sufficient reason, because risks are involved in radiating unborn members of any species. Skeletal development in the fetus is adequate to show on a radiograph at forty-two to forty-five days post-breeding, but fifty-five to fifty-eight days is a better time because the bones are larger and better mineralized.

# Secondary Signs of Pregnancy

Secondary signs of pregnancy can help support your pregnancy diagnosis but alone are not enough to determine that a bitch is or is not in whelp. Some of these signs can be seen to a marked degree of development during pseudopregnancy.

## Enlargement of the Abdomen

Before four weeks post-breeding, a bitch with an average-sized litter probably will not have any detectable swelling of the abdomen. The uterus at this time is not really significantly enlarged unless the litter is quite large. At about five weeks it would be normal to see or feel some fullness, or some loss of the normal tuck-up of the abdomen. In long-haired bitches you will see the hairs that cover the flank fold stick slightly outward instead of downward. After six weeks, enlargement is steady and progressive and of course will vary greatly with the size of the litter. The most growth seems to occur between six and eight weeks post-breeding. Little change is seen during the last week of gestation.

## Haircoat

Nothing particularly related to the coat occurs during pregnancy except that many bitches seem to have an especially luxurious bloom. Much of the hair on the abdomen, around the mammary area, will be shed around forty-five to fifty days post-breeding. The shed can be quite dramatic and is a preparation for nursing. The phenomenon is, of course, much more noticeable in long-haired breeds than others.

Nearly all bitches will shed their entire coat profusely after having a litter, especially breeds with a double coat. The shed begins any time after whelping and is complete by the time the puppies are three months old. This postpartum effluvium, as it is called, is more extreme than any shed seen at any other time in a female or a male, and the bitches will become quite bald. Good nutrition and management will ensure a healthy new coat, which will be complete by the time the pups are seven months old, but nothing can prevent the shed.

## Appearance of the Vulva

Immediately after being in heat, the vulva will be enlarged, although not as swollen as a week or two previously, during estrus. It is one of the less reliable signs, but typically the vulva in a pregnant bitch will remain larger, more relaxed, and more elastic than in one that is not pregnant. It isn't especially significant, however, whether the vulva in a particular individual is tight or loose. For example, a tighter conformation will not necessarily be more inclined to lead to a difficult whelping. What really matters during whelping is the amount of relaxation that occurs and the capacity of the tissue to stretch (elasticity).

## Vaginal Discharge

A clear discharge of mucus will almost always be seen with pregnancy, and it seems to be quite a reliable sign. It is first noticed about four to five weeks after breeding and continues until parturition. To my knowledge, the source of the mucus has never been investigated. It could originate from mucous glands of the cervix.

The function of the mucus is also uncertain. I have seen it in quite copious amounts in some bitches, and in long-haired dogs it tends to dry around the vulva, causing the hairs to stick together in clumps. The dried mucus, by the way, is easy to brush out of the coat, and brushing is certainly the method to remove it, as opposed to bathing or trying to wipe it away before it dries. The mucus should be absolutely clear or only slightly cloudy, and odorless. Any traces of creamy, yellowish, or green color, or strong odor, are signs of infection.

Occasionally you will see some reddish or brownish (old blood) mucous discharge during diestrus, even as late as the fourth week. While this may appear frightening, especially if you are hoping that the bitch is pregnant, it is probably just a bit of residual vaginal or uterine blood that is late in being discharged. If the colored discharge is odorless and clear (not pus-like or creamy), then it is normal. I once noticed a reddish-brown mucus in one of my own bitches at five to six weeks post-breeding. The pregnancy was otherwise perfectly normal. By chance when the bitch whelped, I noticed that a small mummified fetus was delivered just before the normal pups. I speculate that the death of the fetus (which died at about five weeks) caused some discharge of bloody fluids through the cervix. It was presumably implanted nearest the body of the uterus, otherwise no blood would have been able to reach the cervix. The normally developing fetuses and their membranes and placentas would have blocked the way.

## Appetite

Appetite changes are just as variable as the temperaments of our bitches during pregnancy. You would have to say that just about anything goes.

First is "Miss chow hound," who never missed a meal in her life and doesn't miss a meal for the duration of her pregnancy. She's the one everybody wants. Then there is the basically good eater who succumbs to a bit of a decrease in appetite for a week or two early in pregnancy, about the time of implantation (third week). Her appetite may return to normal and remain so until she whelps. Another type of bitch will experience a drop in appetite at the third or fourth week, as above, but never really return to normal until lactation begins. I have known bitches that will not eat even their normal ration of food during an entire pregnancy, much to the dismay of their poor owners. Happily, these noneaters usually produce a normal litter, somehow managing to give enough of their meager intake to their pups. Yet another type of bitch will have a much better appetite than normal, and these, of course, are easy, because their intake can be regulated to suit our wishes.

Nausea is sometimes seen during the second to the fourth week of pregnancy. I can only assume that this is a phenomenon similar to "morning sickness" in women. It doesn't necessarily occur in the morning, but it is a true nausea and occurs in bitches that otherwise seem perfectly normal.

Since the various changes in appetite are so numerous and cover such a wide spectrum, it really isn't possible to say with much confidence what a change means in an individual until her own special pattern has made itself known. If a bitch has morning sickness around three weeks and then seems perfectly normal, for example, there is a good chance that she will do the same thing in a subsequent pregnancy.

## Mammary Development

Every bitch will have some increase in mammary tissue following every heat. This is a result of the hormonal situation during diestrus. Thus, the presence of some mammary enlargement until four weeks post-breeding cannot be used to predict pregnancy. In fact, its presence is evidence that ovulation has occurred during the heat, because progesterone is responsible for much of the change.

After four weeks, mammary development will continue, the nipples will gradually enlarge, and milk can usually be squeezed from the nipples during the final seven to ten days of pregnancy. In a bitch that does not have much of a pseudopregnancy when she is not bred, i.e., who does not develop mammary tissue beyond the expected amount in early diestrus, the continued development later is a good sign of pregnancy. Unfortunately, in bitches that experience pseudopregnancy, development after the fourth week is not a reliable sign of pregnancy.

## Body Temperature

A normal dog, male or female, has an average rectal temperature of 101.5 ° F., with a range of 101 ° to 102.5 ° generally considered to be normal. During the latter stages of pregnancy, the bitch's body temperature will be lower than normal, averaging 100.5 °. Table 1 shows weekly temperatures taken in a group of Beagles during pregnancy. You can see that the decline is gradual and steady, beginning after breeding and reaching the lowest pre-whelping level at term. Non-pregnant bitches may also experience a temperature decline during the first half of diestrus, but their temperature returns to normal while the pregnant bitch's continues to drop during the last trimester. This phenomenon is seldom discussed or observed, because in most cases there is no real need to take a healthy bitch's temperature. Just be aware of the pattern, however, so that if a temperature is taken in late pregnancy, you will not be alarmed or worried about the lower-than-normal level, nor think that it is a sign of impending labor.

**Table 1. AVERAGE BODY TEMPERATURE DURING DIESTRUS.**
Ten pregnant Beagles are compared with ten non-pregnant Beagles.

| Day of Diestrus | Body Temperature (of) Pregnant | Non-pregnant |
|---|---|---|
| Estrus | 101.33 | 101.32 |
| 7 | 101.23 | 101.48 |
| 14 | 101.16 | 101.34 |
| 21 | 101.12 | 101.12 |
| 28 | 101.00 | 101.11 |
| 35 | 100.80 | 100.98 |
| 42 | 100.74 | 101.03 |
| 49 | 100.52 | 101.26 |
| 56 | 100.42 | 101.20 |

At the onset of labor, the bitch's temperature will drop at least a full degree, to 98° to 99.4°. This drop will be seen in virtually all bitches and is the most reliable sign that whelping is about to occur. As a general rule, the first puppy will be seen within twenty-four hours of the temperature drop. The exact mechanisms responsible for the low basal temperature during pregnancy have never been explained, to my knowledge. The pre-whelping temperature drop is caused by the sudden decline in progesterone that occurs before whelping.

## Urination

During late pregnancy, the enlarged uterus, which lies above the urinary bladder, occupies a large part of the space in the abdomen. The uterus puts quite a bit of pressure on the bladder, so that a typical pregnant bitch will need to urinate more frequently than normal.

## Blood Count

The white blood cell count will usually be elevated in a pregnant bitch between twenty-one and forty days, after which time it gradually returns to normal. The elevation averages 7,000 above normal for the bitch. Hemoglobin will be below normal during the last half of pregnancy, and platelets increase, as does the sedimentation rate.

The elevation in white cells, decreased hemoglobin, and an increased sedimentation rate and platelet count, if present, do not positively diagnose pregnancy, because other things can cause the same changes. But if these changes are not present around thirty-five to forty days post-breeding, the bitch probably is not pregnant.

# Vaccination During Pregnancy

It is generally recommended in professional veterinary guidelines for immunization that bitches not be vaccinated with live virus vaccines against distemper, hepatitis, parvovirus, or rabies during pregnancy. The recommendation originated when vaccines were new to the veterinary field and the earliest vaccines were still quite virulent, not as highly modified as they are today. Abortions, reabsorptions, and weak or dead puppies were among the problems that occurred with early distemper vaccines. Modern vaccines, however, are safer and probably would not cause any difficulty. As a matter of safety, you should at least avoid vaccinating between the fifteenth to fortieth days post-breeding, the period during which all of the organ systems are being differentiated and the embryo is sensitive to insult (D12 to D30 when D1 has been determined). In a case where a bitch has not been vaccinated and may be exposed to distemper or parvovirus, it would be better to take the risk involved in vaccination in order to protect her against the disease.

*There is no evidence that parvovirus vaccines interfere in any way with reproduction.* It is important to keep every bitch's vaccination current, for her protection as well as her puppies!

# Prenatal Development

It has always fascinated me to know just what is happening inside a bitch between the time she is bred and when she whelps. The following table of events should help summarize these important events:

| Average Day Post Breeding | Day of Diestrus | Stage of Development |
|---|---|---|
| 1 | D-7 | In estrus, pre-ovulation. |
| 2 | D-6 | Ovulation, primary oocyte. |
| 3 | D-5 | Primary oocytes have moved to distal portion of oviduct, meiosis is occurring. |
| 4 | D-4 | Secondary oocytes ready for fertilization. |
| 5 | D-3 | Fertilization occurs. |
| 6 | D-2 | First cleavage division of ovum begins. |
| 7 | D-1 | Two-cell embryo in oviduct. |
| 8 | D1 | Four-cell embryo; going out of heat. |
| 9 | D2 | 8-16 cells in a cluster. |
| 10 | D3 | 16-32 cells, a morula. |
| 11 | D4 | 32-64 cell morula enters the uterus. |
| 12-18 | D5-11 | Blastocyst forms, floating free in uterus. Individual blastocysts become spaced evenly along horns of uterus. Grow from ¼ mm to 2 mm in diameter. |
| 19 | 12 | Implantation begins. Uterine response in a bandlike area around each embryo, suggesting the shape of the placenta that is beginning to develop for each embryo. |
| 20 | 13 | Embryo body begins to take form as a linear thickening of cells, the primitive streak. |
| 21 | 14 | Primitive body segments, or somites, begin to take form. |
| 22 | 15 | Ten or more somites have formed. Primitive brain and spinal cord growing. Head portion of embryo bends forward. |
| 23-25 | 16-18 | Limb buds form, embryo grows in length from 5 to 10 mm. |
| 26 | 19 | Limb buds paddlelike, 11 mm. Swellings along uterus at site of each implantation may be large enough to palpate for pregnancy detection. |
| 27 | 20 | Eyes begin to have pigment in retina, body taking form, face beginning to take shape. |
| 28 | 21 | Facial features more defined, individual nipples are seen. 14-15 mm in length. |

| 29 | 22 | Toes take shape, whisker buds are visible, umbilical cord forms. 18 mm in length. |
| 30-32 | 23-25 | Differentiation of detail of face and limbs until facial features are definitely doglike. 22 mm in length; 1 gram (1/30 ounce) in weight. |
| 33 | 26 | Male/female sexual characteristics can be differentiated. 24 mm in length. |
| 34 | 27 | Digits and claws are differentiated. Eyelids, which are at first open, close. 30 mm in length; 2 grams in weight. |
| 35-40 | 28-33 | Growth and differentiation of external and internal details. Reaches 45 mm in length and 6 grams in weight. Skin pigment begins to be visible. |
| 41-63 | 34-58 | Further growth and differentiation of the fetus. |

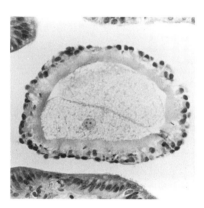

**Fig. 14.2 Two Cell Embryo in the Oviduct.** *(Thin section). About 7 days post-breeding. D-1.*

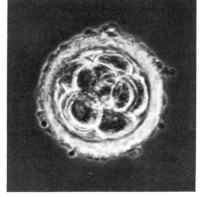

**Fig. 14.3 Morula, Approximately 8 Cells.** *(whole embryo). About 9 days post-breeding. D2.*

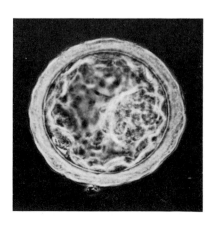

**Fig. 14.4 Early Blastocyst,** *free-floating in the uterus. About 13 days post-breeding. D6. (Whole embryo).*

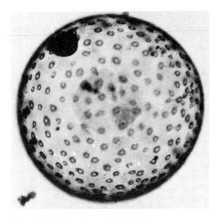

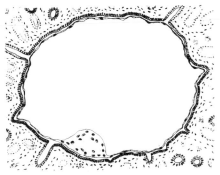

**Fig. 14.5 Blastocyst, Still Free-floating** in the uterus. About 17 days post-breeding. D10. (Whole embryo).

**Fig. 14.6 Blastocyst at About** 19 days post-breeding. D12. The embryo consists of an inner cell mass which will differentiate into the fetus, and a thin layer of trophoblast cells which will become fetal membranes and placenta. The embryo fills the lumen of the uterus and attachment begins.

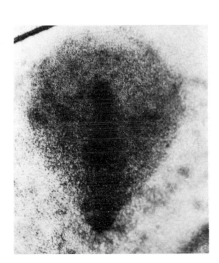

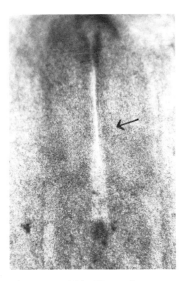

**Fig. 14.7 Primitive Streak Stage Embryo.** Immediately after attachment begins, the inner cell mass flattens and begins to develop the basic body symmetry, a head end, a tail end, with bilateral symmetry. (Whole embryo). About 20 days post-breeding. D13.

**Fig. 14.8 Within Hours the Embryo Has Enlarged, Elongated, and the Beginnings of the Nervous System** (neural groove) form. The first somite (body segment) is barely visible (arrow). (Whole embryo). About 20½ days post-breeding. D13-14.

It is interesting to note that during the first nineteen days or so after breeding, the embryos float freely in the oviduct and uterus and are almost totally undifferentiated. During this time, they seem to be quite resistant to insult, such as chemicals, or irradiation, which might at a later time cause death or malformation.

The period of embryogenesis or organogenesis lasts only about two weeks. It is during this time that the undifferentiated mass of cells in the blastocyst develops all of the various organs and systems of the body. By the end of this period the embryo has become recognizable as a dog and is called a fetus. During the period of organogenesis, the embryo is very sensitive to a variety of insults. Various birth defects, such as cleft palate, omphalocoel (open abdomen), spina bifida (open spine), or defects of the limbs, tail, or skull may occur at this time.

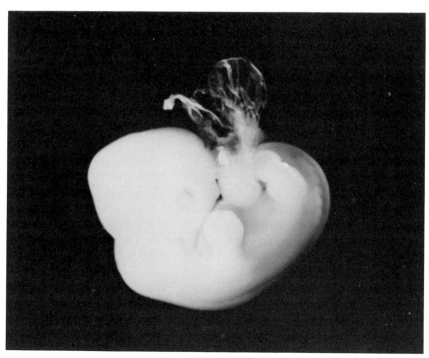

Fig. 14.9 Embryo at Approximately 28 Days Post-Breeding. D21.

After the period of organogenesis, the fetus undergoes refinement of all of its organ systems and is once again fairly resistant to influences that may interfere with normal development.

# The Placenta

The term "placenta" originates in Greek and means "flat cake," an obvious reference to the human placenta.

The placenta is a temporary organ that acts as the link between the developing pup and its mother. It is an intricate meshwork of vessels and supportive tissue, one side maternal and the other side embryonic/fetal, which lie so closely together that nutrients and waste products can readily pass in either direction. Oxygen, nutritional substances, and antibodies and other substances are transferred to the fetus, while urea and carbon dioxide are carried away.

The design or actual shape of the dog's placenta is determined very early, at the time implantation begins, approximately eighteen days post-breeding. At that time, each embryo is an elongated hollow ball of cells (blastocyst) with a shape resembling the GoodYear blimp. The blastocysts lie within the lumen of the uterus and are of sufficient size that there is physical contact between the endometrium and

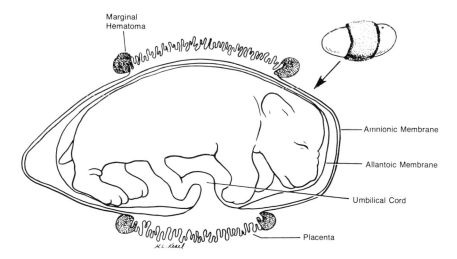

**Fig. 14.10 The Placenta is a Belt-Like Structure Which Encircles the Fetus** and is attached to the uterus around its entire circumference. Contact of the implanting blastocyst determined this configuration at the earliest moments of the placenta's development. A marginal hematoma, dark green in color, is a prominent feature of the canine placenta.

the embryo in a band which completely encircles the largest portion of the blastocyst's circumference. Simultaneously, as the two structures come in contact, both sides begin to change, proliferate, interdigitate, and become intimately connected to form the placenta. The shape of the placenta is bandlike and is technically referred to as "zonary." A napkin ring would be a commonly recognized object that is identical in shape to a canine placenta. As the embryo grows and develops into a fetus, the placenta also grows until, in a full-term fetus, it encircles the pup and all its membranes like a belt.

The edges of the placenta consist of a slightly thickened zone of hematoma (masses of blood that have escaped vessels), which is dark green in color. The green color is mainly a pigment, biliverdin, which is one of the breakdown products of hemoglobin from dead or degenerating red blood cells. The blood in the marginal hematoma originates entirely from the maternal circulation and is believed to provide nourishment, including iron, by its breakdown.

Some hormones are synthesized by the placenta, including progesterone. In some animals, sufficient progesterone is produced to maintain the pregnancy in its later stages. In the bitch, however, placental progesterone is relatively insignificant, progesterone from the corpora lutea being essential to the maintenance of pregnancy throughout its duration. Enzymes are also synthesized, at least some of which are important in the attachment of the embryonic cells to the maternal portion of the organ.

To understand how such an intimate relationship can exist, we need to look a little deeper into the actual structure of the placenta on a microscopic level. At first, imagine the maternal side—the endometrium. It is made up of a layer of epithelium which lines the lumen, plus connective tissue beneath the epithelium, plus capillaries, the small ends of the uterine blood supply. The embryonic side consists of the same three layers. The epithelium is derived from the outermost layer of the blastocyst and is referred to as the trophoblast. There is also a layer of connective tissue, and capillaries which connect the organ to the fetal blood supply via the umbilical vessels.

In some animals, including the horse, the maternal and fetal sides of the placenta retain all three cellular layers. The close apposition of the two sides is enough to allow transfer between the two circulations. In the dog, however, two of the original six layers are lost. The trophoblast cells (embryonic epithelium) are so aggressive that they erode the maternal epithelium and connective tissue down to the maternal capillaries, leaving the capillaries in direct contact with the trophoblast. Whether this more intimate relationship has any real advantage over that seen in other animals is uncertain. A mare can certainly support a large fetus with its more loosely applied placenta.

Chapter Fifteen

# PUPPIES ALWAYS COME IN THE MIDDLE OF THE NIGHT: WHELPING

## Length of Gestation

The length of gestation in bitches is quite variable when measured from the day of breeding. The average is sixty-three days, with a range from fifty-nine to seventy days. The wide range results from the fact that breeding can occur during such a long time around ovulation and be successful. Bitches bred early in their cycle, before ovulation, will seem to have a longer gestation than those bred after ovulation. In the former group, there was a delay before actual conception and embryonic development could begin. In the later groups, development began almost immediately.

Gestation length is much less variable when the onset of diestrus is used to calculate the length. On that day, all bitches' pregnancies are at the same stage, whether they were bred three days earlier or ten days earlier. When diestrus is used, the gestation length averages fifty-seven days, with a high percentage of bitches (85 percent) whelping within one day of the average. The onset of diestrus (D1) is three days after the actual day of fertilization on D-3. The actual length of a bitch's gestation, then, is sixty days, counting from the day of conception. Making vaginal smears daily during the last few days of the cycle and determining D1 is extremely valuable in deriving an expected whelping day.

The litter size can have an influence on whelping day, larger-than-average litters being whelped in many cases earlier than expected, and unusually small litters later than expected. This pattern is much more obvious when diestral timing is used (Table 1).

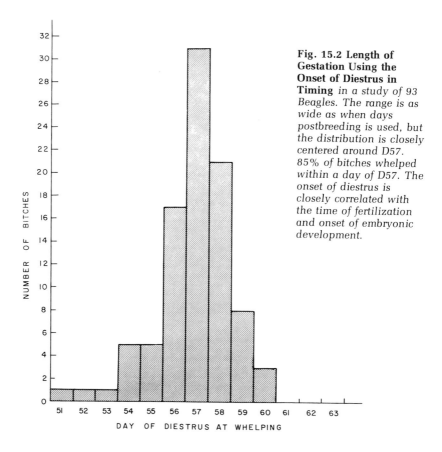

**Fig. 15.1 Length of Gestation (Days Postbreeding) in a Study of 74 Beagles.** *The wide range and fairly even distribution result from the fact that breeding during a long time span before fertilization and onset of embryonic development is successful in the bitch.*

**Fig. 15.2 Length of Gestation Using the Onset of Diestrus in Timing** *in a study of 93 Beagles. The range is as wide as when days postbreeding is used, but the distribution is closely centered around D57. 85% of bitches whelped within a day of D57. The onset of diestrus is closely correlated with the time of fertilization and onset of embryonic development.*

Table 1. Gestation Length Related to Litter Size. A Comparison of Timing Gestation From Breeding and From Onset of Diestrus.

| Length of Gestation | | | | | | |
|---|---|---|---|---|---|---|
| Days Postbreeding | | | Day of Diestrus | | | |
| 59 | (2)* | 5.5 pups per litter | D54 | (5)* | 5.2 pups per litter |
| 60 | (4) | 5.4 pups per litter | D55 | (5) | 6.9 pups per litter |
| 61 | (7) | 5.9 pups per litter | D56 | (18) | 4.6 pups per litter |
| 62 | (12) | 5.3 pups per litter | D57 | (32) | 5.0 pups per litter |
| 63 | (11) | 5.1 pups per litter | D58 | (22) | 4.8 pups per litter |
| 64 | (13) | 5.0 pups per litter | D59 | (11) | 3.2 pups per litter |
| 65 | (11) | 3.5 pups per litter | D60 | (7) | 2.2 pups per litter |
| 66 | (11) | 4.0 pups per litter | D61 | (2) | 1.0 pups per litter |
| 67 | (4) | 4.2 pups per litter | D62 | (3) | 1.0 pups per litter |
| 68 | (4) | 4.2 pups per litter | | (108) | |

(80)

* ( ) = Number of litters in each group.

Adapted from Am.J.Vet.Res. Vol 35, No 3, 1974. Reprinted with permission.

# Hormonal Triggers of Whelping

What is it that triggers the complex chain of events we call parturition? At some moment in time, the uterus, which was until that moment content with its contents, begins to object and evacuate itself. The cervix, which was tightly closed, opens, softens, and dilates to allow expulsion to take place.

Progesterone levels during the last days of gestation are slightly elevated, and approximately one day before whelping they drop to low basal levels. Just how important this drop is remains undefined. It is known that progesterone from the corpora lutea is essential to the maintenance of pregnancy in the bitch. The placenta does not assume a significant role in the production of progesterone, as is the case in some other mammals.

Corticosteroids rise dramatically during the final day before labor begins, coincidental with the drop in progesterone. High levels of steroids given during late gestation have been shown to cause abortion, or the onset of labor. Steroids are normally secreted in situations of stress, but the nature of the stress at term pregnancy is uncertain. And the source of the steroid is at this time unknown. Perhaps it is maternal—a response to the distension of the uterus or hormonal changes. Perhaps the source is fetal—a response of the fetuses to crowding, hormonal changes, or stimuli of which we are still unaware.

Estradiol rises a few hours to a day before parturition in some species, but a change in estrogen level related to parturition has not been observed in bitches.

# Stages of Parturition

The first stage of labor is when the cervix dilates. The earliest sign of impending labor in a bitch is usually a decrease in body temperature caused by the abrupt drop in progesterone levels before whelping. If the temperature is measured every twelve hours, you can almost always observe the change. During the last days of pregnancy, the bitch's body temperature is subnormal anyway. It will range from $100^8$ to $100^2$ degrees F. when taken at rest. The drop that signals labor is $98^0$ to $99^4$ degrees F. It is a fairly good rule of thumb that the first puppy can be expected within twenty-four hours of the temperature drop. You can anticipate some variation, of course, when you are only measuring every twelve hours. The longest I have had one of my own bitches go was thirty hours from temperature drop to delivery of the first puppy.

Other early signs of first-stage labor include loss of appetite or poor appetite. A bitch that had been eating well may refuse a meal, yet act otherwise quite normally. Many bitches will seek a place to hide and seclude themselves. Some will go to great lengths to build a nest, shredding papers or bedding, and rearranging it by pushing it around with their noses. This stage may last for many hours, and it is during this time that the cervix begins to dilate. Presumably the bitch has labor pains or discomfort from the uterine contractions during this stage, and if you carefully feel her uterus through the abdomen, you may be able to feel its intermittent tension and relaxation.

The next aspect of first-stage labor involves slowly but steadily increasing discomfort. The bitch will rest at intervals and then pant, move around, change positions, perhaps lie on her back, and generally appear uneasy. This stage may last for several hours, and the periods of uneasiness and restlessness will occur more frequently as the hours pass.

Second-stage labor is when the fetuses are delivered. If you are watching carefully, you will be able to detect the first labor contractions signalling the onset of this second stage. The bitch will display an abdominal press; she may or may not make a sound, such as a groan or grunt. Usually she will appear to have an urgent need to eliminate. She may rush to the door and ask to go out if she is a house pet, or she may have a bowel movement and/or urinate. It appears that the sensation of a puppy entering the birth canal is very much like that of a bowel movement; at least the bitch seems to interpret it that way. My bitches at home usually ask to go out several times during this stage. They squat as if to eliminate but after the first time do not pass much, if any, feces or urine.

Vomiting is a variable behavior during early second-stage labor. If the bitch has not eaten for more than twelve hours, which is the usual pattern, only a small amount of stomach fluid will be brought up. Of course, if she is one of those true "chow hounds" that didn't miss a meal, the last meal may be lost. Just why some vomit and others do not is uncertain. It is probably related to the general situation of pain and apprehension that the bitch is experiencing at the time.

The labor contractions at first are observed at intervals of ten or more minutes. They usually come in waves of three to five, followed by a rest. The bitch may sit up, lie on her side, assume a squatting position, or even stand and arch her back. She may purse her lips as if to push harder, or there may be little evidence of her conscious straining. At this time, a puppy is present in the birth canal, and its presence, which puts pressure on the dorsal wall of the vagina, stimulates the hardest labor contractions. As the pup moves down the canal, the contractions may become stronger and almost always will become more frequent. A discharge of mucus, presumably the cervical mucous plug, will be passed before the first pup. It is not always seen, and whether or not it is observed is of no consequence.

As the pup reaches the vulva, its outer fluid-filled sac, the amnionic sac, pushes through the vulva and will usually be the first structure you will see. Alternatively, the sac may rupture and the fluid be discharged before you see any part of the puppy. The rupturing of the fluid sac does not put the puppy in jeopardy. The only complication related to rupture of the sac would be if it ruptured early and the delivery was delayed long enough for the vaginal lining to become dry. The sac and its fluids are important in lubrication. The pup and its inner membrane, the allantoic sac, should be delivered during one of the labor contractions. In the ideal situation with an average-size pup, a well-lubricated vagina, and a fully relaxed vulva, the entire pup will be delivered in one to three contractions. If any of those conditions are less than ideal, the delivery will be more difficult. The bitch may have to strain harder, the pup may become stuck half in and half out, or it may not be able to start through the vulva without assistance.

When the puppy is presented at the vulva, the bitch normally begins to lick the area quite intensely. She will lick up the amnionic fluid, lick herself, and lick the puppy in almost a random manner. Her licking serves to break the allantoic membrane which immediately

138

**Fig. 15.3 Second Stage Labor.** *The first structure normally seen protruding from the vulva is the amnionic sac filled with fluid. When the sac is ruptured a discharge of fluid will be seen.*

**Fig. 15.4 The Puppy, Still Surrounded by the Amnionic Sac, is Delivered.** *In many cases the sac will have ruptured by this time.*

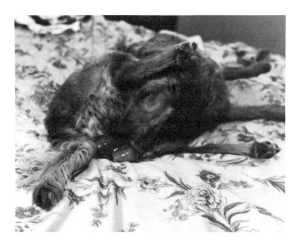

**Fig. 15.5 The Bitch Has Ruptured the Amnionic Sac by Licking** *and the allantoic sac, which closely surrounds the puppy, is seen.*

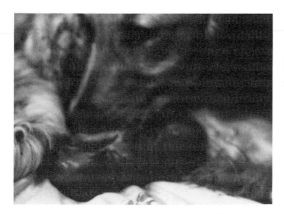

**Fig. 15.6 Continued Licking Has Removed the Allantoic** *sac, and the pup has taken its first breath. The placenta remains in the birth canal and the pup's umbilical cord is still intact at this point.*

**Fig. 15.7 The Fetal Membranes and Placenta are Being** *pulled out of the vagina by the bitch. She will normally eat the placenta immediately and chew the umbilical cord, crushing it in the process, which prevents bleeding.*

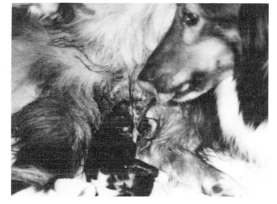

covers the pup, if it was not already ruptured during the delivery. The licking stimulates the puppy and removes fluid from around its head, facilitating its breathing. The bitch will normally be very intent upon disposing of the fetal membranes, eating them immediately. If the placenta is delivered along with the puppy or shortly after, she will grab it and swallow it immediately. At virtually the same time, she will begin to chew the umbilical cord with her cheek teeth— not the incisors. The chewing seems to crush the cord and pinch off the vessels so that bleeding under the natural situation is not a problem. The length of the cord will vary from quite short to a couple of ragged inches. Excess length seems to bother most bitches, and they continue to gnaw at the end until it finally suits them. It is not unusual for the bitch to lift the puppy off the ground while working on the cord. This in itself will not cause a problem, such as umbilical hernia; in fact, it is believed that pulling on the umbilical cord of a newborn pup stimulates its breathing via a nervous system feedback.

When the placenta is not delivered along with the puppy, the bitch licks the cord, still lying in the vulva, or pulls at it until it is presented. If the cord breaks, nothing can be done except to wait and watch for that placenta to be passed later in the delivery. Either posterior (rear feet first) or anterior (head first) presentation is normal, with approximately 60 percent of pups being anterior.

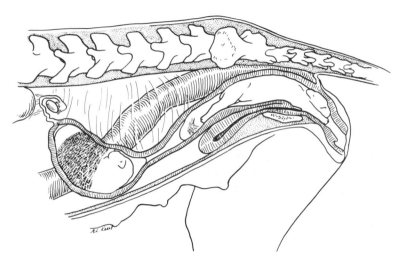

**Fig. 15.8 Anterior Presentation.** *Each pup must be lifted over the rim of the pelvis, then pushed downward through the vagina. Approximately 60% of deliveries are anterior and 40% posterior. Uterine contractions are localized around the caudal-most fetus, leaving the others relatively undisturbed.*

The interval between puppies varies from a few minutes to several hours. It is normal to rest between delivery of individual pups, with the length of the rest having no real significance. The average would be about thirty minutes. It is known that pups are delivered randomly from the two horns, rather than from one horn until empty and then the other horn. It might seem logical for one horn to empty, then the other, since many bitches deliver part of the litter, rest for quite a long time (up to eight hours), then finish. But I have not seen that pattern at surgery in bitches that have delivered part of their litter, and one study in which seven-week fetuses were marked with radioactive injections and later observed as to their whelping order supports this understanding.

After the final pup has been delivered the bitch should seem more relaxed, more comfortable, and usually very tired. She will tend to the puppies and sleep for most of the next day. In large breeds or any individual with a large litter, the final pup may be delayed. It seems that the uterus does not maintain the pace of the delivery and labor is stopped. It is not unusual for the final puppy or two to be delivered after a five- to eight-hour rest, or even longer.

The fluids eliminated during whelping are normally clear (amnionic fluids), to blood-tinged (placental detachment, bleeding), to dark green (placental marginal hematoma pigment). Large amounts of bright blood may indicate excessive hemorrhaging from a detached

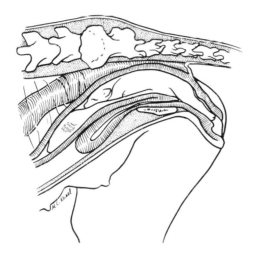

Fig. 15.9 **Posterior Presentation** *is normal and seen in approximately 40% of deliveries.*

placental site. Usually further labor and uterine contraction will stop the bleeding. Large amounts of green pigment are an indication of partial or total detachment of a placenta, with breakdown, and may or may not indicate serious trouble. In the case of dystocia (difficult birth) because of large fetuses, small pelvis or malpresentation, the detachment, if complete, will result in death of the fetus due to suffocation. If the detachment is only partial, the fetus will be all right, at least for a while.

The discharge of dark blackish green pigment usually means a long-standing detachment of a placenta with advanced degeneration of the placental pigments. It should be assumed that a fetus has died. The dead fetus and its membranes will be delivered during the course of the whelping if the labor is otherwise normal. The presence of a dead fetus per se is not cause for alarm; it seldom endangers the healthy pups. The reason for the detachment may be a real concern— other fetuses may be in jeopardy from the same problem.

A thick dark brown discharge passed during whelping is an indication of a fetal death and earlier reabsorption. If examined closely, the material may contain a structureless solid mass, the remains of the dead fetus or a mummified fetus. This situation does not endanger the whelping or the healthy fetuses.

# Signs of Trouble

First-stage labor, the period of cervical dilation, is prolonged. It probably begins at the time of the temperature drop. If labor has not begun within thirty hours of the temperature drop, the bitch should be examined by your veterinarian. The most common reason for failure to start labor is primary uterine inertia, or a lack of tone in the uterine musculature. I would suspect hormonal abnormalities but know of no studies that have investigated this possibility.

From the time the first visible contraction is seen, the first puppy ought to be delivered within two hours, although up to three hours can be normal. It is the passage of the puppy into and through the vagina that stimulates these contractions. A gloved finger inserted into the vagina and pressed against the dorsal vaginal wall will stimulate labor contractions in the same way as a pup in the passage. The technique is called feathering. The hardest contractions are stimulated by the presence of a pup in the caudal vagina. If almost continuous, hard contractions persist for more than thirty minutes without delivery, there is likely to be trouble. Either the pup is too large to make the final exit through the vulva, or it may be malpositioned.

Prolonged second-stage labor may be caused by disproportionately large pups or malpresentations, such as the back presented first or forelegs extending into the body of the uterus with the head extending up one horn and rear quarters up the opposite horn. Poor uterine tone can also be a cause, in which case the muscular wall does not develop contractions strong enough to move the puppies along. Too small a maternal pelvis for the size of the pups is another possibility. This would include, of course, constrictions of the pelvic canal following trauma and pelvic fracture. In this case, the pup would be trapped in the body of the uterus, unable to pass into the vagina. Weak to moderate and intermittent contractions would probably be seen, without the normal progress to the strong and frequent voluntary contractions that immediately precede delivery.

# Criteria For Dystocia

A call to the veterinarian is indicated when:

1. More than twenty-four hours have passed since the temperature drop without signs of second-stage labor (thirty hours maximum).

2. Mild or intermittent labor contractions have been seen for more than two hours without progressing to hard labor (three hours maximum).

3. Hard, almost continuous contractions have been occurring for more than thirty minutes without appearance of the puppy at the vulva.

4. Head, nose, or rear legs and tail protruding through the vulva for more than fifteen minutes and you are unable to pull.

5. More than six hours rest (no labor) between puppies, and you know that she is not finished.

# What to Do in Case of Trouble

### Determining the Problem

The first step is to determine if there is indeed a problem. More than three hours in intermittent labor is a positive indication that you should try to determine what is going on. If labor has been constant and vigorous, because of the presence of a pup in the vagina, more than thirty minutes is an indication for examination.

The more inexperienced breeder should seek the services of the veterinarian; those with more experience might be able to do some things at home. A digital examination with a well-lubricated, gloved finger will often tell the source of the problem. A poorly dilated birth canal or vulvar area can easily be determined. The presence or absence of a pup in the canal is very important to know. If a pup can be felt, is it anterior or posterior? Does there seem to be adequate moisture for lubrication or is the canal dry? How large does the pup seem to be? To me, a pup in the birth canal always feels huge, and I am constantly amazed at how inaccurate my perception seems to be when I am feeling with the tip of one finger. Does the pressure of your finger on the dorsal vaginal wall stimulate labor? If so, does the labor succeed in moving a puppy further into the vagina? A pup far enough back may be grasped and pulled. This must be done gently, pulling not straight backward, but in an arch corresponding to that of the vagina rising over the pelvis and then dropping downward through the vulva. This is a technique best left to the veterinarian unless you have had extensive experience. It is sometimes extremely difficult to pull a pup, especially if it is disproportionately large for the bitch's birth canal.

## Episiotomy

Another means of assisting a bitch, although not actually used often, is an episiotomy. This can be useful in the case of a pup that has passed the entire length of the vagina and is lodged at the vulva because of inadequate relaxation of the vulvar tissue. The procedure should only be done by a veterinarian. A vertical incision is made with surgical scissors through the dorsal wall of the vulva and perineal tissue, in the direction of the anus. After delivery, the veterinarian will suture the incision. Healing proceeds quickly because of the excellent blood supply to the area.

## Cesarian Section

In cases of dystocia (abnormal delivery) in which either the life of the bitch or the lives of the puppies are in danger, a C-section, the surgical delivery of the litter, is the solution. There are a number of absolute indications for a C-section:

1. Maternal causes.
   a. Small or collapsed pelvic canal.
   b. Inadequate relaxation, dilation of the perivulvar tissue.
   c. Primary or secondary uterine inertia.
2. Fetal causes.
   a. Fetus too large for the size of the bitch. This can include abnormal pups, monsters.
   b. Fetus malpositioned, making it impossible to enter or pass through the vagina.
   c. Two fetuses lodged at the body of the uterus, not allowing either one to enter the vagina.

Other factors may be involved in making the decision to do a section. One may be a great desire to deliver the puppies alive—an unwillingness to risk losing any of them. The delivery of a valuable litter of purebred pups can and often is an entirely different proposition than delivering a batch of unplanned and unwanted mongrels. In the latter case, for example, you might opt for attempting to pull a malpositioned puppy, thus risking its loss, in order to allow the normal delivery of the ones to follow. That risk may be totally unacceptable in a valuable purebred litter.

# Oxytocin

Injections of pituitary oxytocin hormone can be useful, even seemingly miraculous, in some cases. The hormone stimulates uterine contractions and delivers the puppies in a normal labor. An injection of oxytocin in a normal laboring bitch will cause strong uterine contractions within a few minutes. That is about all oxytocin can do—stimulate strong uterine contractions of a normal uterus in a laboring bitch. It is certainly not the answer to every case of abnormal or difficult labor. Consider the bitch that is having difficulty because of a small pelvis or a large puppy. The puppy may be killed by the contractions because it simply will not pass through the birth canal. The same is true of a malpositioned puppy. Oxytocin in such a situation will probably result in at least one dead puppy. A bitch with a fully dilated cervix, puppies that are properly positioned, and a fatigued uterus, may be helped with oxytocin. It can be useful in helping delivery of the last puppy of a large litter. But a word of caution is advisable here. It is normal and natural for the bitch to take a prolonged rest during delivery of a large litter. If oxytocin is given, it may cause not only the next puppy to be moved down toward and through the vagina, but may also result in detachment of its placenta, along with the placentas of following pups, and a rupture of their amnionic sacs and loss of fluid, and ultimately result in delivery of puppies dead from suffocation.

Oxytocin should never be given simply to speed up delivery. (See Criteria for Dystocia #5.) It should never be given to stimulate the start of labor. It should not be given unless a digital examination by a veterinarian has been done to determine pelvic size, cervical dilation, and presence and position of the presenting fetus. The more dystocia cases I see, the less inclined I am to reach for the oxytocin vial. The

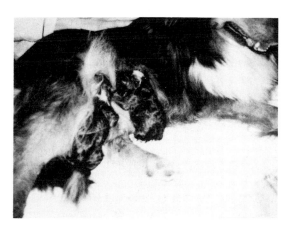

**Fig. 15.10 The Suckling of the Newborn Pups Stimulates Oxytocin Release,** *which in turn stimulates further labor. The pups should be left with the bitch as much as possible in order to make labor as natural and as short as possible.*

suckling of newborn puppies stimulates a release of oxytocin via a nervous system feedback mechanism. The oxytocin stimulates the letletdown of the bitch's milk, as well as contractions of the uterus. So in a normal situation it is best to leave the newborn pups with the bitch and allow them to nurse. In so doing, her labor will be stimulated and the rest of the litter will be delivered as quickly as possible. It is normal and natural for this to occur, and I believe that many people make a mistake in grabbing the puppies away from the bitch in order to warm them under a heat lamp, or whatever. I have seen bitches go into labor immediately when the pups were returned to her and started suckling. Normal pups will not be hurt by a little prolonged time of dampness if they are in the whelping area during delivery. The benefit that they get from early nursing and the help that they are to the bitch are far more important. The pups can be removed temporarily as each new one is delivered.

Following delivery, the bitch should be examined by a veterinarian. If there is any doubt about the delivery of the final pup or pups in a large litter from a large bitch, a radiograph could be taken. In most cases, it is possible to tell by palpation of the uterus whether deliver is complete. If the uterus is properly contracted and has good muscular tone, no injections or treatment of any kind are indicated. It is not necessary to give oxytocin routinely after every delivery. Such treatment is simply not needed in a normal bitch with healthy whelps that are suckling normally. On the other hand, if the delivery has not been normal (if, for example, the bitch had uterine inertia and had to have puppies pulled, or had premature detachment of placentas, retained placentas, or other complications), an injection of oxytocin to stimulate uterine regression would be in order.

Antibiotic injections and infusions are not routinely needed either in a bitch that had a normal delivery. In fact, there is a potential for doing more harm than good with an injection because of the possibility of creating digestive upsets in puppies after they nurse and ingest the drug in the bitch's milk. If there has been a difficult delivery and/or digital examination of the vagina, or retained placentas, an injection or infusion of antibiotic may be indicated to help prevent future infection. In such a case, the potential for good would outweigh the risk of problems.

# Helping the Newborn Puppies

Normal healthy newborns will not need our assistance in getting a good start in life. Their mother will provide them with what they need, i.e., stimulation, a clear airway, warmth, nourishment, and detachment from their placenta at the umbilicus. Unfortunately, not every puppy is normal and not every bitch is the perfect mother. Every newborn needs gentle stimulation. The bitch will lick them, roll them from side to side, nuzzle them, and pull on the umbilical cord. All of this stimulates their muscular activity immediately after birth, most importantly breathing. If the bitch ignores her pup, you are well advised to help out by picking it up in a soft towel and rubbing it, massaging all parts of its body and limbs until it moves around and is breathing well.

There will always be some fluid and mucus in a newborn puppy's mouth. The bitch will normally remove this by licking. If she doesn't do it, you must help by swabbing the pup's nostrils and mouth with a towel. A rubber bulb syringe, the type designed to flush ears, is ideal for aspirating fluid from the pup's throat. Hold the puppy with head lowered to allow gravity to assist in draining fluids from the airway. In cases where there seems to be a lot of fluid and the pup is having difficulty breathing, detach the umbilicus (see below), then lie the pup in your hands with the head held securely between your fingers. Raise the pup at full arm's length above your head, and swing your arms in a semicircle downward toward the floor. This quick motion will help force fluids from the airway and assist breathing.

There is no need to be in a hurry to force the newborn into immediate vigorous movement with a lot of crying. In many cases, the pups need a few moments of rest to regain some strength expended during delivery. Be sure that the pup is breathing regularly and that he will respond by moving to your physical stimulation, and let him rest.

The bitch will generally be as concerned with the umbilical cord as with the puppy, licking the vulva vigorously and trying to get hold of the cord and pull the placenta and membranes out. Do not be overzealous in cutting the umbilical cord; the puppy is perfectly safe while still attached and if anything may still be able to retrieve some of its own blood from the fetal side of the placenta. If the bitch shows no interest, gently grasp the cord, preferably with a towel, and gently pull and try to withdraw the afterbirth. The cord may break, but it is better if it does not so that the placenta can be accounted for after the delivery of each puppy. Small hemostatic forceps can be clamped to the cord and used to help withdraw the placenta. Don't rush; it takes some time for the placenta to be loosened from the uterus enough to be removed.

If the bitch shows no inclination to chew the umbilical cord, you may need to sever it. But do so only after the entire afterbirth has been delivered. Grasp the cord between your fingers with a towel and tear the cord in two, or alternatively grasp between two hemostats and pull the cord in two between the forceps. The tearing simulates the natural method of crushing and chewing and helps seal the vessels. Avoid cutting with a sharp scissors. This will almost always result in bleeding. When bleeding occurs, tie off the umbilicus with a strand of sewing thread about one-half inch from the pup's body. Antiseptic solutions are seldom needed.

The bitch may be allowed to eat each afterbirth. This is normal and natural. A bit of controversy exists about whether to allow the bitch to consume the afterbirths, and if so how many. My opinion is that since eating the afterbirths is a natural thing for her to do the bitch ought to be allowed to do as she wishes. It may cause vomiting and a soft stool to be passed the following day, but this is a relatively insignificant problem. My common sense tells me that there must be a tremendous amount of nourishment in the afterbirth. Why waste it?

As soon as the pup is detached from its afterbirth, it can be dried with a towel, examined for sex, checked for obvious defects (such as cleft palate or limb or tail abnormalities), and either placed in a warm area or back with the bitch, depending on her willingness to care for it at the moment. It is best to leave the pups at the mother's side as much as possible for reasons explained previously. Early nourishment is important but not essential for survival. Weaker puppies will require some rest before they have gained enough strength to nurse without help. The strongest may start to nurse immediately before their umbilical cords are even detached and afterbirths delivered. It is fairly safe to wait twelve hours or even more before the pups nurse, and in some of the larger litters, that much time may easily elapse before the entire litter is delivered. When the final puppy is delivered and the bitch finally is able to relax and begin taking care of the new family, watch closely and see that every puppy is nursing. Help the small or weak ones by holding them up to the nipple and keeping them stimulated so that they don't fall asleep before taking some nourishment. It is important that every puppy ingest some of the bitch's first milk, or colostrum, within twenty-four hours of whelping. Colostrum contains antibodies that protect the puppies during their first weeks against diseases such as distemper and parvovirus. The pups' ability to absorb antibodies from the intestines disappears after twenty-four hours.

After a cesarian delivery, the pups will often require a little extra care. The stress of surgery and pain will usually delay the development of the bitch's normal mothering behavior. Special care must be taken to see that the puppies stay together close to the bitch and that they are all nursing. Be prepared to supplement their feed for two to three days. It usually takes that long before normal lactation is sufficient to keep the pups full. Whether it is a delay in milk production or in letdown is uncertain.

A bitch that is spayed during the C-section can be expected to have a perfectly normal lactation. If it is your intention to spay her for any reason, don't hesitate to have it done at the time. It will spare her the stress of another surgery later.

# Postpartum

The bitch will have a discharge after whelping, and it is referred to as "lochia." The lochia should be reddish to reddish-brown, (green-tinged is normal the first day), almost odorless, mucoid, and fairly clear. Thick creamy or heavy greyish discharge would be abnormal and a sign of infection, as would a foul odor. The lochia may have a dark green color in the case of a retained placenta. Retained placenta per se is not dangerous. It will normally degenerate and be eliminated. It may, however, serve as an added source of nourishment to bacteria in case of infection. The lochia will persist in decreasing amounts for four to six weeks after parturition. Ideally it should stop in about three weeks, but in my experience, normal bitches may have some discharge for longer. Poor involution of the uterus may be responsible for persistent discharge, especially if it is blood-tinged.

Occasionally a bitch will seem nervous or take no interest in her pups during the first day or so, will not keep them close to her in a group, and will not want to stay with them. Two things might help such a bitch. Oxytocin can be given to stimulate milk letdown, which may help settle the bitch as the pups suckle. Progesterone can also be helpful in some cases, presumably because it has a mild tranquilizing effect on the nervous system. Progesterone should be used conservatively because it interferes with the release of prolactin.

# Cannibalism

One of the most distressing problems that can be encountered is a bitch that mutilates or eats her newborn puppies. In many cases it may simply be accidental—the bitch may bite the umbilical cord too close and evisceration of the puppy may occur. This could happen in bitches with poor occlusion of the teeth. The bitch apparently does not have any sense of where to stop chewing and does not realize that she is hurting the puppy. Some puppies are born with defective closure of the umbilical area (omphalocele) with an open abdomen, and this may go unnoticed in the flurry of activity that occurs at whelping. Later the bitch's licking may eviscerate the puppy. In some cases cannibalism seems to be caused by a "mental disorder" in the bitch—some kind of psychological derangement. Such a bitch should not be bred as she is likely to do the same thing again.

**Chapter Sixteen**

# NATURE'S PERFECT FOOD: LACTATION

Lactation is the process of production and delivery of milk from the glands of the mother to newborn animals. Milk production for the newborn is such a fundamental process of life that the whole mammalian class is defined on the characteristic of possessing mammary glands. Dogs, being mammals, are no exception.

A mammary gland is a highly modified gland in the skin. A dog of either sex has four to six pairs of glands, usually five pairs, located in two parallel rows along the ventral abdomen. The nipples or teats are the structures through which the glands empty. The nipples are developed quite early in prenatal life and can be seen about four weeks post-breeding in the embryo.

It is not unusual to see a missing nipple or pair, or asymmetrically placed nipples in some individuals. Most deviations from the five perfectly spaced and placed pairs are of no consequence. Things have a way of working out, as they say, and even in cases where the number of puppies is more than the number of teats, all should be able to survive by establishing a rotation system of feeding. Even pups in smaller litters often feed in shifts, some sleeping while others feed.

A mammary gland has a structure basically similar to other glands that secrete a fluid, such as salivary or tear glands. The secretion, in this case milk, is produced by cells within the gland and carried to the outside by a system of ducts. The secretory cells are arranged in groups lining small sacs called alveoli. (Air sacs in the lungs are also called alveoli.) The secretory cells produce milk, and the alveoli fill and stretch. Groups of alveoli are arranged in clusters called lobules, and the milk is drained from the lobules through small ducts. A cluster of eight to twenty-two large ducts drains to the outside at each teat. Anybody who has ever milked a cow knows that its teat is emptied by one large central duct. In the dog it is quite different, the outflow being more like a sieve, making milking a dog very difficult.

Before a young bitch's first heat, there is little development of the mammary glands. Estrogen, which is produced mainly during pro-estrus, is primarily responsible for development of the duct system.

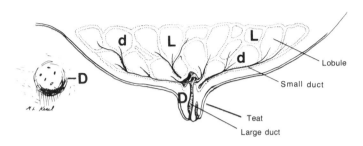

**Fig. 16.1 Structure of the Mammary Gland During Lactation.** *Milk is produced in the alveoli, which are arranged in clusters, or lobules (L) separated by connective tissue. Small ducts (d) drain the alveoli and join as large ducts (D) which empty through the teat. The sieve-like arrangement of ducts exiting the teat is shown.*

Progesterone, produced primarily during diestrus, is responsible for proliferation of the secretory tissue. Prolactin and growth hormone, two pituitary hormones, are also involved.

During pregnancy, the development of the mammary tissue proceeds in roughly three stages. The duct system and vascular bed (arteries and veins) proliferate in the early part of pregnancy. Alveoli and lobules develop during the middle to latter part of pregnancy. The growth that is seen during the latest stages is due to dilation of the alveoli with milk. The filling with milk accounts for most of the actual size of the glands at the end of pregnancy.

During pregnancy, the actual production of milk is inhibited by estrogen and progesterone. At the time of whelping, progesterone drops, removing the inhibition to milk production. Prolactin and ACTH rise at this time and probably contribute to the beginning of lactation.

The amount of mammary development and the amount of milk produced at the time of whelping do not appear to be correlated with the number of pups in the litter. Every bitch seems to be prepared to nurse a good number of puppies, and if there are only a few, or if the pups do not survive, some adjustment must be made.

Maintenance of lactation depends on the presence of pups that are suckling and upon the release of prolactin, ACTH, and oxytocin. These hormones are released as a response to suckling. Suckling also obviously removes milk from the alveoli so that more can be produced. The exact function of prolactin and ACTH in lactation is not clear. Oxytocin, however, is better understood. The suckling stimulates the glands physically and causes nervous system stimulation, which leads

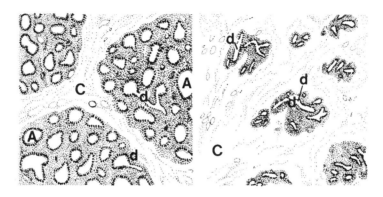

**Fig. 16.2 Microscopic Structure of Lactating and Non-Lactating Mammary Glands.** *During lactation alveoli (A), small collecting ducts (d), and large ducts (not shown) are well-developed. Connective tissue (C) separates lobules. In the non-lactating gland only the duct system and connective tissue are present.*

to release of oxytocin from the pituitary. Oxytocin in turn causes muscle fibers in the alveoli to contract and ducts to widen, thus ejecting milk from the alveoli. This process is called "letdown of milk," and facilitates suckling. Oxytocin also causes contractions of the uterus, aiding in its return to normal size after whelping.

Suckling has another effect on the central nervous system, and that is a stimulation of the water and food intake control centers of the hypothalamus. The bitch has an increase in appetite which enables her to eat the extra nutrients needed to support the lactation.

If there are no pups to suckle, or if there are only a few, the glands that were prepared for full lactation will regress quite rapidly. Failure to remove milk from the glands causes a buildup of pressure in the alveoli, which in turn causes a cessation of secretion. The decrease in secretory cells and degeneration of the alveoli occur quite rapidly.

The first few days after whelping can be a time of concern for some bitches. In some cases the mammary glands fill to the point that they seem quite hard and uncomfortable. This can happen in a short time, such as when the bitch habitually lies only on one side and the lower glands are not emptied properly. It is also often seen when the litter is small. Without the pressure of competition, most newborns will not crawl under the bitch's rear leg and use the inguinal pair of glands. These glands, then, may become excessively full and hard. Within a couple of days, the necessary adjustment will be made, and the unused glands will decrease their activity. They can be gently milked if necessary to relieve some pressure, but care must be taken to avoid removing so much that further production is stimulated. Food should be withheld for twenty-four hours to decrease milk production. A veterinarian may need to give diuretics in some severe cases.

# Milk

Milk is high in calories and balanced in nutrients. The exact composition varies among different species of mammals, depending on the special needs of the young of that species. The composition of bitch's milk is compared with the cow and the goat, two common sources of supplemental feed for puppies:

|  | Dog | Cow | Goat |
|---|---|---|---|
| % Fat | 8.3 | 4-5 | 3.5 |
| % Protein | 9.5 | 3.8 | 3.1 |
| % Lactose | 3.7 | 5.0 | 4.6 |
| % Ash | 1.2 | 0.7 | 0.79 |
| % Total Solids | 20.7 | 14-15 | 12.0 |

The percentages of fat, protein and total solids are higher in dogs' than in cows' and goats' milk.

Colostrum is the first milk secreted after birth, and its composition is slightly different from the milk that follows. Generally, colostrum is higher in percent solids, protein, and ash, while lower in percent lactose. The most important feature of colostrum is that it contains antibodies that are transferred to the newborn pups during the first twenty-four hours of life. The antibodies protect the pups against many diseases until they are several weeks of age. Without colostrum, many more newborn animals would be lost to simple infections during the early days of their lives.

# Acid Milk

Breeders have come to me with a complaint that puppies are not doing well because of acid milk. This is not a disease or syndrome that I have been able to recognize. It usually means that the milk is inadequate and the pups are not thriving. Or it may be that the bitch is ill and her metabolic problems have interfered with lactation. It has been shown that hydrochloric acid can be added to milk with no adverse effects on the puppies, so acid per se is not a problem. Normally the pH of milk is 6.3 to 6.9, which is slightly acidic. There is acid in the stomach of all animals, and milk will curdle immediately after being ingested. A puppy with an indigestion problem may vomit some curdled milk. This may be interpreted as acid milk, but this is a puppy problem rather than a problem with the milk itself.

Chapter Seventeen

# PUPPIES, PUPPIES, PUPPIES: PEDIATRICS

## Normal Newborn Puppies

Surveys have shown that as many as 30 percent of newborn puppies fail to survive to weaning age. This may happen to us if we as
breeders adopt a strict "hands off" attitude, if we never interfere with
or assist any puppy under any circumstances and do not attend our
whelpings. But I for one cannot do that. When I have a small or weak
pup, or a normal pup in trouble, I simply have to help it. I don't
believe that we need to lose a third of our puppies. Surely we've
worked too hard to produce them to be willing to settle for that. Many
puppies that might have died without our help can grow up to be
normal, strong members of their breed. We should be able to keep
the loss of puppies from birth to weaning below 10 percent.

The first step in managing a litter of newborn puppies is to
recognize a normal, healthy pup. Several characteristics are important to recognize, and once you have experienced a healthy litter, you
will undoubtedly be quite familiar with the signs even though you
may never have verbalized or itemized your observations. In other
words, you'll know that a puppy is normal simply because it looks
and acts normally.

Puppies are relatively immature at birth. Their eyelids are sealed
shut, so they are blind. Their ears are also sealed, so they are deaf.
The eyelids open at ten to fourteen days in most litters, with some
as late as eighteen days. It is a temptation to say that puppies that
open their eyes later are "younger" at birth than the ones that open
earlier, but nothing definitive can really be concluded from the age
of eye-opening because there is considerable normal variation. Ears
open a few days later than the eyes, usually at thirteen to seventeen

days of age. The umbilical cord should become dry and hard within hours of birth and will fall off at two to three days of age.

A normal puppy has good tone. Immediately after birth, actual muscle tone is not exceptionally good, but healthy puppies seem to gain strength and vitality continuously. When you pick up a puppy, it should wiggle in your hands with a lot of energy. It should feel firm and plump. When a healthy puppy moves its legs to crawl, it gets somewhere (assuming the bedding is adequate to give good traction). It will crawl forward or in a wide circle rather quickly, with head weaving from side to side, seeking contact with its dam and littermates, and seeking warmth. It will sound off with intermittent sharp cries until it has reestablished contact with something warm.

Healthy puppies will usually be quiet. If they are warm enough they will seldom cry, even though they may be hungry. When they are hungry, though, they may cry and be restless and fussy. After ten to fifteen minutes, they will become tired and fall asleep. Excessive crying may have a more serious cause. Pups may cry when the bitch's milk does not agree with them—an unusual problem. Colic, septicemia, or viremia may also be involved. With these latter problems, other signs of trouble will be seen in addition to the pups' crying.

A normal puppy will feel warm to the touch. During the first six days, puppies are unable to shiver, and they are very poor at regulating their own body temperature. They rely on heat from the bitch's body, especially the mammary area, and from their littermates to maintain their body heat. During the first two weeks, the pups' rectal temperatures will be 94 to 97 degrees F. From two to four weeks it should range from 97 to 99 degrees F., and then range from 100 to 101.5 degrees F. after four weeks of age. The pups' skin temperature seems to be important for the bitch to recognize and respond with normal maternal behavior to the pups. Chilled puppies are often neglected by the bitch. Skin temperature may not be the only reason, however, that a bitch will cull a chilled pup. A chilled puppy will not be able to move, suckle, and respond with its normal behavior, thus breaking the bond of recognition between the bitch and the puppy.

Puppies are never really still while asleep. They normally will lie in a sprawled position, on their sternum or their side, and they are continuously jerking and twitching, stretching and shifting their position. Ears twitch, lips pucker, feet jerk, and heads jump constantly. This pattern is called "activated sleep," and it is important to the development of the neuromuscular system. It seems to be the mechanism whereby the newborn pups develop muscle tone and coordination. Pups experience activated sleep during 75 percent of their sleeping time. Beware of the puppy that lies quietly while asleep!

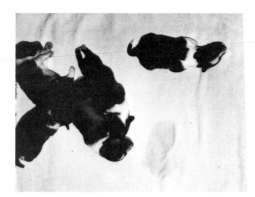

Fig. 17.1 Activated Sleep *is normal and essential to development of the neuro-muscular system. Compare with figure 2, taken about two minutes later.*

Fig. 17.2 A Sick Puppy is Too Quiet. *At about two minutes four of these puppies have stretched, twitched, or shifted position. One lies quietly, legs withdrawn, head extended, and has not moved (upper right). This puppy was suffering from pneumonia and did not survive.*

Certain reflexes are present in newborn puppies and can be checked to be sure that everything is normal. The most important reflex is suckling. When the puppy's mouth feels the teat, it should immediately begin strong suckling activity. The most vigorous pups will begin suckling within seconds of their birth, but others need some time—up to several hours—before they begin. It is not unusual for some newborns to seem quite ineffective about finding a teat and beginning to feed, but a normal pup will accomplish this on his own, in time. Healthy pups will compete vigorously for a teat.

A reflex or behavior closely related to suckling is the rooting behavior. A normal puppy will push forward with its head and crawl upward when a hand or other object is placed in contact. This behavior helps the pups locate teats. When placed on their back, normal puppies will immediately right themselves. This is a response to gravity, or an innate response to the sense of which way is up. Newborn pups also have some reaction to odors, to pain, and to touch and have a wink reflex to touch around the eyes, even though the eyelids are closed.

**160**

Fig. 17.3 A Strong Suckling Reflex is demonstrated by this healthy two day old puppy.

Fig. 17.4 The Rooting Reflex, pushing forward and upward when pressure is applied around the face, probably aids in locating the teats. It is seen during the first three weeks.

Fig. 17.5 A and B The Righting Reflex enables the pup to remain in a normal upright position. When placed on its back, the pup immediately rolls over, legs extended.

Normal puppies will be extremely clean, as will their bedding. Cleanliness, of course, depends on the instinctive behavior of the bitch to keep her nest clean. Given a normal bitch in this regard, signs of fecal material on the pups or bedding usually means diarrhea and possible trouble. Puppies depend on stimulation from the bitch to urinate and defecate during approximately the first sixteen days. The

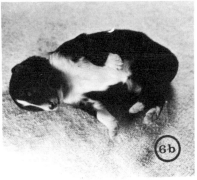

**Fig. 17.6 A and B. Normal Postures For a Two Day Old Puppy.** *Newborn puppies tend to curl, head and legs drawn inward because of dominance of the flexor muscles. Older puppies tend to extend head and limbs more.*

bitch's licking stimulates elimination, and this is one mechanism that serves to keep the nest or whelping box clean. Gradually the pups gain the ability to eliminate without stimulation, and after four weeks reflex elimination cannot be elicited.

Weight gain is perhaps the single most important indication of how the puppies are doing. It is normal for every puppy to make a steady slight weight gain every day, even the first day. They should double their birth weight by seven to ten days. Pups that gain daily are doing well. A second category of pups may drop a little weight (no more than 10 percent in the first twenty-four hours of life), then start to gain, and they likewise will be fine. A third category of pups, the ones that lose more than 10 percent of their body weight in the first twenty-four hours, have a poor survival rate unless they get some supplemental feeding. A good rule of thumb is that a healthy puppy can be expected to gain 1 to 1¼ grams for every pound of expected adult body weight per day during their first three to four weeks (30 grams = 1 ounce). I have often seen even faster gains than this (1½ to 2 ounces per day for Shelties, which will mature at twenty to twenty-five pounds). It is extremely important to have a scale and to weigh puppies regularly. A household scale that will weigh in ounces is adequate. A shoebox makes a nice platform to hold pups for weighing if your scale does not have a tray. Weight should be taken and charted daily for the first two weeks, then every two to three days for another two to three weeks. Puppies have little body fat at birth and therefore little stored energy. Their energy is stored mainly in the form of glycogen in the liver and heart and can be rapidly depleted if they are not getting enough nourishment. Glycogen stores will last about twenty-four hours if no nourishment is being taken in.

Finally, a few vital statistics for newborn puppies, compared with adults:

| Respiration: | First day: | 8-18 breaths/minute |
| | To 5 weeks: | 15-35 breaths/minute |
| | Adult: | 10-30 breaths/minute |
| Heart rate: | First day: | 120-150 beats/minute |
| | First 14 days: | 180-220 beats/minute |
| | 2 to 5 weeks: | 150-220 beats/minute |
| | Adult: | 80-140 beats/minute |

Body consists of 82 percent water at birth, 68 percent water after 5 months.

Learn to recognize a healthy puppy, and you will be able to spot a sick pup immediately. A newborn puppy can slip from normal to desperately ill or chilled in only a few hours. It is extremely important, then, to check every puppy several times a day and watch for a few minutes to observe their suckling behavior, their response to their dam entering the whelping box, and their activated sleep. Every puppy should be picked up twice a day to feel its tone, warmth, and strength and to check for soiling under the tail. A sick pup will not act, look, or feel right. It may be limp or cool to the touch, or it may be lying with legs sprawled outward, head turned to one side with little activated sleep activity. It might not suck your finger like the others, or it might crawl away from the pile of littermates and fall asleep. It won't gain weight and it won't feel warm.

# Neonatal Care

When your litter is healthy and the dam has abundant milk and mothering instinct, there is little you need to do for them. The single most important thing they need is a warm environment, especially the first two weeks. Much of this warmth comes from the dam's body. To insure adequate warmth inside the whelping box, two things can be done. Cover the box, for one. A blanket over the top is ideal. It is natural for a dog to feel secure in an enclosed area; they were den-living creatures originally. Second, provide some external source of heat, such as a heating pad or a heat lamp. A heating pad should be covered with a towel or blanket and can be kept on "low" at all times. When the bitch leaves the box, the puppies can crawl onto the pad if they need the extra warmth. A heat lamp must be carefully placed and adjusted to prevent overheating. In either case, the heat source should be placed to the side or end of the box so that the bitch is not forced to lie on or under it all the time. She may not need or appreciate the extra warmth herself.

Clean bedding and privacy are about the only other requirements if everything is going normally. The pups need a surface on which to lie that will give them traction when nursing. Newspaper is not adequate because its surface is too smooth. It is ideal, however, to place underneath other bedding because of its absorbency. Towels, blankets, mattress pads, or any kind of washable rugs or other material will do. Some bitches have a tendency to scratch at their bedding, and if the bedding is loose, puppies might be buried. A folded mattress pad is ideal in such a situation.

The bitch should not be locked in with her litter in a small area, such as a crate, so that she cannot get away from the puppies. Granted, most bitches will refuse to leave their litter during the first few days, but in later days they need to be able to get away from them. A whelping box divided into compartments or a crate that is left open (with some kind of raised barrier at the door to keep the pups inside), or a box that the bitch can step out of, are perfectly acceptable.

# Routine Surgery

Many breed standards require that tails be docked, and some that rear dewclaws be removed. Usually the removal of front dewclaws is optional but highly desirable because of general appearance, ease of grooming, and prevention of future injury in the field. These routine procedures are best done when the puppies are two to five days of age. The exact age can vary somewhat depending on how big and vigorous the puppies are, whether they are gaining weight as desired, whether they happen to reach the average three days old during the middle of a show weekend or at a more convenient time. If delayed beyond ten days or so, anesthetic should be used, and since it is difficult and risky to anesthetize very young puppies, surgery should be delayed until about the time of weaning or beyond.

Many breeders are skillful at these minor surgical procedures and are able to do their own tail and dewclaw amputations. Many techniques exist, of course, and probably many are satisfactory. The method I prefer for dewclaw removal is to clamp with a small curved hemostatic forcep along the base of the dewclaw, parallel and flush with the surface of the leg. After several seconds, the clamp is removed

**Fig. 17.7 A and B. Dewclaw Removal Begins With Clamping the** *dewclaw with a small hemostatic forcep. The clamp is placed parallel and close to the leg, both for front and rear dewclaws.*

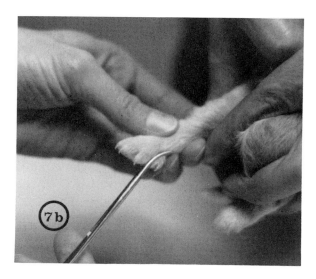

and the dewclaw cut off with a scissors. Clamping will usually reduce bleeding to an acceptable level, but silver nitrate cautery sticks are always at hand to cauterize the ones that continue to bleed. There is no need to dissect into the leg to amputate at any particular level of the toe's bony structure. Sutures in the skin are also not needed. A scab will form within a day and will fall off after about two weeks, leaving a small but truly inconspicuous scar. Any swelling around the scab, moisture, or pus with looseness of the scab are signs of infection and should be treated immediately by a veterinarian.

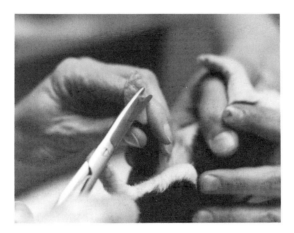

Fig. 17.8 After Clamping For Several Seconds, the Dewclaw is Cut With a Scissors *flush with the leg. No further dissection or sutures are required.*

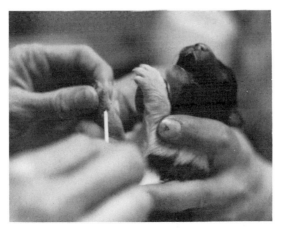

Fig. 17.9 Silver Nitrate Cautery Sticks are Used to Control Bleeding *if it occurs. The stick is placed firmly on the wound and held for a couple seconds. Oh, yes, it does hurt.*

Tail docking can be accomplished with a number of different techniques, too. The one that I use is to cut the tail with scissors in a U-shape to the desired length (i.e., the arms of the U toward the pup's body, the closed end away), and to place a single suture across the cut end. The technique leaves a rounded shape to the tip of the tail and it heals nicely.

All pups should have their toenails trimmed every week, starting with the very first week of their lives.

Fig. 17.10 Every Puppy Should Have a Nail Trim at Least Once a Week.

# Orphan Puppies

Puppies that have lost their dam or that must be raised by hand for any reason are a challenge; they require a great deal of attention and commitment. Raising orphan puppies, though, is rewarding and should be successful if certain guidelines are followed and the needs of the puppies are kept foremost in mind. Research scientists routinely raise puppies by hand for experimental puposes (such as to produce pathogen-free animals for testing vaccines), and the art of orphan puppy care is well-developed.

The needs of orphans are the same as those of any newborn: warmth, food, stimulation to eliminate, cleanliness. They do not have the bitch's body as a source of heat, so their environmental temperature must be maintained at a higher level. The puppy can only maintain its body temperature about seven degrees above the room temperature. Ninety-degree ambient temperature is desirable during the first five days. A higher temperature (95 degrees F) will overheat the pup, and it will have a bright red skin color because of dilation of skin vessels. The temperature can be gradually decreased to 80 degrees F during the second and third weeks, to 75 degress F during the fourth to sixth weeks, and then to 70 degrees F or normal room temperature by the eighth week.

Feeding to achieve normal daily weight gains is important. The pups should be weighed at least twice a day for their first four days, then daily until three weeks. After three weeks, once or twice weekly weights should be recorded until weaning. The puppies should be able to double their birth weight in seven to ten days, just as any pup being raised naturally. They need seventy cubic centimeters of water every day per pound of body weight.

Two types of feed for newborns are available, commercial and homemade. Prepared commercial bitch replacement formulas are excellent. Results are good and it is convenient. Home-prepared formulas can also be used, but they do cost in terms of time and inconvenience. Here is a commonly used formula for newborn pups:

1 cup whole milk
Yolk of one to three eggs
Few drops of infant vitamins, especially vitamin E
1 tablespoon corn oil

This formula provides approximately 1.6 calories per gram, which is equal to one milliliter or one cubic centimeter (five milliliters equals one teaspoon; fifteen milliliters equals one tablespoon.) In order to figure the requirements for daily feeding of each puppy, use the following caloric requirement table as a guide (from *The Basic Guide to Canine Nutrition*):

| Age in Weeks | Cal. per lb. body weight per day |
|:---:|:---:|
| 1 | 60-70 |
| 2 | 70-80 |
| 3 | 80-90 |
| 4 | 90-100 |
| over 4 | 100 |

A sample calculation: How much formula would be needed for an eight ounce puppy at less than one week? This pup is one-half pound so would need at least thirty calories. Each milliliter gives 1.6 calories, so $\frac{30 \text{ cal.}}{1.6 \text{ cal./ml.}}$ = 18.75 ml. formula. This is equivalent to 3.75 teaspoons, or rounded, approximately four teaspoons. Commercial bitch's milk formulas are excellent as well, such as *Esbilac®*, *Ortholac®*, or *Unilac®*. *Esbilac®* has approximately one calorie per cubic centimeter of mixed formula.

When feeding a commercial formula, prepare and feed according to the manufacturer's directions. When fed the amounts needed to supply the above caloric requirements, the puppies will not be fat, but they will be able to handle the feed without developing diarrhea, and they will grow adequately. Goat's milk can be used fairly successfully. It is easier to digest than cow's milk, even though the amounts of protein and fat are about the same.

Puppies need to be fed small amounts at frequent intervals. Feeding every two to three hours is best during the first couple of days, then four to six times per day until three weeks. From three to six weeks, four feedings daily are enough. After each feeding, it is important that each pup is handled, massaged, or rubbed, to burp it. Elimination of urine and feces should be stimulated by stroking with a warm, moist cloth or tissue.

Orphan puppies will not have their need to suckle satisfied in a normal way, and because of this they should be kept separated from one another. If left together they will suck on each other's tails, legs, and genitals, and this may cause skin irritation. The need to suck may be fulfilled by switching to bottle feeding at about nine days of age, by providing a suitable object as a pacifier.

There are two satisfactory methods to feed the puppies—by bottle or by stomach tube. An eye dropper is not satisfactory. It is too difficult to control the amount given with a dropper and too easy for the pups to aspirate into their lungs. The pups do not have a good gag reflex during their first few days and are not stimulated to swallow adequately by milk simply dropped into their mouths. The procedure is too slow as well.

Bottle feeding can be satisfactory, though it tends to be time-consuming, especially with large litters. For toy breeds, a doll-size bottle and nipple can be used, with a human-size better for large breeds. The holes in the nipple must be of adequate size to release the proper amount of milk. Too fast a flow may make the pups choke, while too little will result in the pups' becoming fatigued before they have eaten enough. The holes should be large enough so that when the bottle is held upside-down, milk slowly drips from the tip. A twenty-one-gauge hypodermic needle is useful to make the holes. Heat the needle to red hot and punch two holes in the nipple.

The safest, easiest, and most efficient method to feed orphans or any puppy that requires supplementation is to use a stomach tube. I recommend this method to everyone because it is so easy and so safe. The equipment needed consists of a syringe and a soft rubber tube, which should be available from any veterinarian. An 8 French urinary/feeding catheter is ideal for pups up to eight ounces, and larger sizes for larger pups, up to 16 French for a one-and-one-quarter-pound pup. The tube is attached directly to the syringe and should have an end that is specially designed to adapt to a standard-size syringe. The tube should be marked to provide a guide as to the correct amount to insert. To mark the tube, measure the distance from the tip of the puppy's nose to the level of its last rib. Mark with tape at that point.

To fill the syringe, simply insert the end of the feeding tube into the pup's formula and draw up a milliliter or two more than you intend to feed. Never mind the air in the tube. It is simply the volume of air that filled the feeding tube and will not be given to the puppy as long as the syringe is held upright with the bubble at the top, and you have a little more formula in the syringe than you plan to feed.

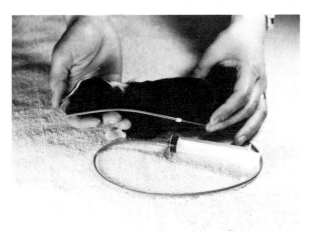

Fig. 17.11 **Feeding Puppies by Stomach Tube** *is convenient, quick and safe. The distance from the tip of the nose to the last rib is marked to aid in determining the correct distance to insert the tube.*

Passage of the tube into the pup's stomach is also simple. Place the tip of the tube on top of the pup's tongue, and slide slowly, steadily, and gently over the tongue to the back of the throat. Contact of the tube on the back of the throat will stimulate a swallow reflex and the tube will continue down into the stomach. Two ways to check that the tube is in the stomach:

1. If the tube is inserted to your mark, it is in the stomach. If by chance it went into the trachea, it would not pass the full length. The trachea ends at the level of about the pup's elbow. Be aware that a small tube in a larger puppy may pass into the lungs, so it is important to use a tube that is appropriate for the pup's size.
2. Check the puppy's breathing. If he is breathing and/or crying, the tube cannot be in his trachea. An older pup would cough and be in severe respiratory distress if the tube was misdi-

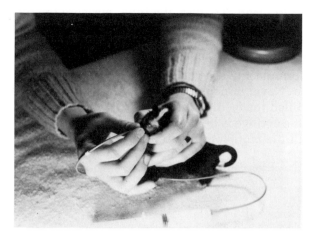

**Fig. 17.12 The Feeding Tube is Slipped into the Mouth,**
*over the tongue, and with a gentle but steady pressure
pushed until it has been inserted to the mark.*

**Fig. 17.13 The Desired Amount of Formula is Slowly**
*injected into the puppy's stomach. The air in the syringe
will not be injected if the air bubble is kept high, as
shown.*

rected. Do not rush. Take a few extra seconds to check for
breathing and to be sure that the tube has been inserted to its
mark. Then you will know for sure that the tube is in the
stomach and can slowly inject the formula without fear.

Your feeding equipment, whether bottles or syringe and tube
should be thoroughly washed between feedings but does not need
to be sterilized.

# Premature Puppies

A true premature puppy is one that is born before its full gestation length, that is consequently lower than normal birth weight, and that shows signs of immaturity. Not every low birth weight pup is premature. It is hard to say at precisely what gestation length a puppy is premature because there is so much normal variation. Some litters may be fully mature at fifty-nine days post-breeding, while others might be premature. The difference would depend on whether the breeding occurred early in estrus, several days before ovulation, or later, after ovulation. In the former case, fifty-nine days could be too early, as the expected gestation might be up to sixty-seven days, but in the latter, it could be the full normal gestation.

A premature puppy is weaker and more easily chilled than a normal puppy. It needs a higher ambient humidity because of a tendency to dehydrate easily. It will have a scanty or poorly developed haircoat. The face and legs and feet especially may seem poorly covered with hair. The legs may look bright pink because of this. The digestive system may not be fully functional, so feeding a premature puppy can be a challenge. Many of them cannot digest milk or milk replacer formula. Their sucking reflex may be weak, and suckling may be ineffective, although they may spend time on the bitch's teats. They should be fed by stomach tube at least until they have grown strong enough for effective suckling. A diet consisting of nutrients that are readily available with a minimum of digestion may be lifesaving. Plain glucose in water can be a helpful supplement at first, at least until the pups have gained some strength.

## Diet for Premature Puppies

| | |
|---|---|
| 5% Amino acid solution | 100 ml |
| 50% Glucose | 75 ml |
| 95% Ethyl alcohol | 4 ml |
| Potassium Chloride | 4 milli equivalents |
| Lactated Ringer's Solution | 15 ml |

This diet provides 0.9 Kcal per milliliter, which is only a little lower than *Esbilac®*. It can be used exclusively for the first week, then mixed 50/50 with milk replacer formula for a while. The percentage of formula can be increased as the pups grow and become able to digest normal feed.

Premature puppies have poor muscle tone and sometimes seem unable to extend their feet forward to crawl. They crawl on the tops of the carpus (pastern) and tend to develop scabs along the front of their forelegs and feet, presumably because the skin, which lacks hair, is abraded. The legs will extend normally and the scabs disappear as the pups grow and gain strength. Premature pups should be left with the bitch as long as she will accept them and mother them normally.

# Weaning

Whether raised by their mother or as orphans, the process of weaning is the same. It involves a gradual change in diet from milk to dog food beginning at three weeks of age. The pups' first food should be the same as they are going to be fed throughout their puppyhood. It must be softened and mixed with water to form a soupy gruel that can be lapped. At three weeks the pups will readily lick from a shallow pan but do not seem to be able to pick up particles of feed or to chew. Keep the bitch away from the pups for a couple hours before and during feeding so that they will be hungry and will not decide to nurse instead of eat. They should be offered food three or four times a day from three to six weeks of age. The consistency of their feed can be made increasingly solid as they grow older. During the last week before weaning is complete, gradually decrease the amount of feed given to the bitch so that her milk production will decrease. It helps to increase the time that she is kept away from the pups, too. This helps to break the pups' habit of nursing and to decrease the bitch's milk production.

There is absolutely no advantage in weaning early or at some arbitrary time, such as at five weeks, either to the puppies or to the bitch. Left to their own devices, most bitches will wean their pups completely by seven weeks, but some will of course be earlier or later. I have always thought it a bit cruel to separate a bitch from her litter at four or five weeks as some breeders insist because of convenience. The pups at that age are still dependent on the bitch and learning from her and it is not natural for her to be finished with lactation. Too sudden a withdrawal from her pups can lead to caked and uncomfortable mammary glands.

Some bitches have retained the behavior of their more primitive canine ancestors and will regurgitate food for their pups around the time of weaning. This appears to be a reflex behavior and is stimulated by the puppies' nursing and jumping around or licking the bitch's mouth. It is perfectly natural, and there is no need to discourage it.

The pups will eat the regurgitated food with relish if allowed to do so. I have seen some bitches who do not seem to be able to carry the behavior to its normal conclusion and insist on eating the regurgitated food themselves, not allowing their pups to have any.

In cases where weaning must be done suddenly or at an arbitrary time before lactation is naturally finished, a reduced feeding schedule can help the bitch to dry up more comfortably:

First 24 hours—no feed
2nd 24 hours—¼ normal maintenance feed
3rd 24 hours—½ normal maintenance feed
4th 24 hours—¾ normal maintenance feed
5th 24 hours—normal maintenance feed.

Chapter Eighteen

# NEWBORN PUPPIES IN TROUBLE

Puppies that get into trouble, either because of disease or trauma, congenital problems, or simple separation and chilling, all will show a similar set of symptoms. In some cases they will cry. If the crying persists for fifteen minutes or more, you should look for the cause. A healthy puppy that cries because it is hungry will stop in ten to fifteen minutes from fatigue and will fall asleep. Sick pups may appear quiet or listless; they will feel cool to the touch, will not nurse effectively or at all, and will lie separated from their littermates. If the puppy is more than three days old, it should lie with its head extended. If the head is flexed forward, chin on the chest, the puppy is probably sick. You may find a sick puppy that was apparently perfectly normal only a few hours before. The first thing to do is to make a careful, thorough inspection of the puppy in an effort to determine the problem. Look at its umbilical area for swelling, discoloration, or discharge. Check its eyes for swelling or discharge, especially if the eyelids are still closed. Check each leg that has had a dewclaw amputated, and the tail if docked. Check the genital area, the anus, the nose for any discharge, the skin for signs of discoloration or bruising, and the mouth for milk regurgitation. Observe the nature of the puppy's breathing.

A chilled puppy will have a low internal body temperature, usually from 78 to 85 degrees F. Its metabolism will be depressed accordingly; its digestive system will be paralyzed. It will often have excessive moisture around the lips. Its heart rate and breathing will be depressed. If the temperature is below 70 degrees F., you probably cannot tell whether the puppy is dead or alive. There is a way to test for signs of life in such a severely chilled puppy. If the pup is dropped from a height of six to eight inches and is alive, it will stretch after hitting the table or floor. Then you will know that it is alive and might be saved.

It is tempting to want to immediately feed a chilled puppy, but this can do more harm than good. Formula or milk will simply sit in the paralyzed stomach. Never feed formula to a pup whose rectal temperature is below 94 degrees F. Its digestive system is totally inactive. It cannot suckle effectively. The pup needs slow warming and a relative humidity of 55 to 65 percent. One of the best ways to achieve slow warming is to hold the pup inside your clothing next to your body so that the pup is warmed gradually by your body's heat. The motion when you move about will be beneficial in preventing congestion in the lungs. When a pup is placed in a warm environment, such as on a heating pad, and heated too quickly, its skin quickly starts to feel warm. But for a time its core temperature is still depressed, its heart and breathing are still metabolically depressed, and the cardiovascular system is not able to supply the oxygen needed to meet the demands of the warmer extremities. The puppy cannot constrict its peripheral vessels to conserve heat until it is a week of age. In the most severe cases, the peripheral tissues suffer damage from oxygen starvation, resulting in death of cells, hemorrhage, and eventually death of the puppy.

All puppies are born with low prothrombin levels, an important clotting factor in the blood. When sick, they can usually be helped by being given vitamin $K_1$ to overcome the deficiency. After about a week of age, enough vitamin K is made by the bacterial flora in the gut so that supplementation is not needed.

The chilled pup's need for energy can be partially satisfied by giving small amounts of sugar, preferably glucose, or honey in water into the stomach by tube. The sugar, unlike milk, can be absorbed directly from the stomach and can help satisfy the need for energy and counteract hypoglycemia. Dehydration can also be a problem in a sick puppy and may also be aided by the feeding of honey, Karo Syrup®, or sugar in water. A teaspoon of sugar to an ounce of water is adequate. Two milliliters every thirty minutes can be given to an eight-ounce puppy until dehydration is corrected. Subcutaneous fluids may be needed and should be given by your veterinarian. One sign of dehydration is the urine's specific gravity, an indicator of concentration of solids. A normal pup will have a specific gravity of 1.006 to 1.017, whereas in a dehydrated pup it may be 1.035 to 1.040, indicating a higher concentration because of inadequate body fluids.

After the sick puppy has been warmed to normal temperature, rehydrated, and given some energy food, its condition can be reevaluated to determine whether any other specific treatment may be needed for the particular problem.

# Fading Puppy Syndrome

The term "fading puppy" is quite a lot like the term "cancer." It is not a disease, but rather a common non-specific name applied to many different diseases. A fading puppy is a victim of one of a number of problems that starts him into a downward spiral of vitality from which he cannot escape without help. Fading puppies generally will die within four days of first showing signs of the syndrome, or after whelping.

Problems that can result in fading puppies are:
- Excessive mothering by the bitch; pups not getting enough rest.
- Low birth weight for breed, less than 75 percent of normal.
- Bitch's nipples too large for effective suckling.
- Physiologic immaturity, secondary to small placenta, crowding in the uterus.
- Poorly inflated lungs, or hyalin membrane disease.
- Birth defects, external or internal.
  Hydrocephalus (can be related to high nitrates in the water).
  Cleft palate.
  Imperforate anus.
  Segmental aplasia of the bowel.
  Heart defects.
  Vascular defects.
  Thoracic flattening, with respiratory compromise. (Commonly called "swimmers".)
  Vascular ring anomalies.
  Defects in any other organ system.
- Immune deficiency.
- Trauma.
- Metabolic diseases.
- Infections.

## Antibiotic Treatment Related to Fading Puppy Syndrome

It has been suggested that treatment of the bitch or the puppies with antibiotics can cause fading puppy syndrome, presumably by sterilizing the gut. A study to test this theory was done using six different antibiotics administered to newborn puppies for about twelve days. Penicillin and streptomycin, ampicillin, oxytetracycline hydrochloride, kanamycin, gentamycin, and chloramphenicol were tested. None of the treated puppies developed fading puppy syndrome, nor did any digestive problems occur. Interestingly, the treated puppies gained weight at a faster rate than the non-treated ones. Livestock and poultry producers routinely give their growing animals and birds antibiotics in order to optimize growth, and it appears that the same phenomenon may occur in dogs. This is not to suggest, however, that antibiotic treatment of healthy puppies be used routinely. They are still best reserved to treat specific infections.

## Viruses and Infections That Cause Fading Puppies

**Peritonitis.** This occurs when bacterial infection localizes in the abdominal cavity. It may originate from contamination of the umbilical cord at birth. Bacterial toxins paralyze the bowel. *Symptoms*: Crying, bloating or splinting of the abdomen from pain, refusal to eat. Fluid may accumulate in the abdomen. Dark discoloration of the groin area. *Prevention*: Sanitation. Prevent infections in the bitch. *Treatment*: Antibiotics, both systemic and locally in the abdomen; supportive care.

**Septicemia.** This is a bacterial infection that spreads through the bloodstream. *B-Streptococcus, Staphylococcus*, and *E coli* are the most common agents. It may originate from contamination of the umbilical cord or of dewclaw or tail docking wounds, or from infections of the bitch. *Symptoms*: Lethargy, refusal to eat, crying, bloating, swelling of the umbilical cord area or the limb. Dry anal area (i.e., no diarrhea). *Prevention*: Sanitation; good ventilation. *Treatment*: Antibiotics, supportive care, including *Lactobacillus sp* culture to aid digestion. Trocarize with a small needle to reduce severe bloating (this should be done by your veterinarian).

**Viremia (Herpes virus).** This is caused by exposure to herpes virus at the time of whelping. The infection may also cause abortion, still-born, or runted pups. A bitch usually develops antibodies, and future litters will not be affected, but it occasional may be seen in subsequent litters. Apparently healthy animals may shed the virus. It affects only newborn puppies during the first three weeks due to their low body temperatures. The virus is temperature-sensitive and will not cause serious problems in animals with body temperatures over 100 degrees F.; it replicates at 95 to 96 degrees F. *Symptoms:* Soft, odorless, yellow-green stool. Depression, refusal to eat, uneasiness, continuous painful crying. Death in twelve to twenty-four hours. Internally: Mottled kidneys and liver, lungs firm, wet, and mottled. *Prevention:* Unknown. *Treatment:* Nursing care. Puppies that have started crying probably cannot be saved and if they survive will have serious kidney problems. Raise environmental temperature to 100 degrees F. in order to elevate pups' body temperature to 100 degrees F. for at least three hours. Then keep them at 95 degrees F. ambient temperature for the rest of the twenty-four-hour day. This will interfere with viral replication. Recovered pups may have kidney or liver damage.

**Canine Brucellosis.** Discussed in Chapter 18.

**Toxoplasmosis.** This is caused by a protozoon parasite that can lead to various internal problems.

**Infectious Canine Hepatitis**

**Toxins** from certain bacterial infections.

# Other Problems Affecting Newborn Puppies

## Toxic Milk Syndrome

This problem originates with poor involution of the uterus and production of toxins that are excreted in the milk. It may or may not involve retention of placentas after whelping. Toxins in the milk result in digestive upset in the pups. *Symptoms:* Diarrhea, green stools, crying, bloating, salivation, reddened anus. The bitch appears normal, but her uterus has not involuted normally. *Prevention:* Treat the uterine problem in the bitch immediately. *Treatment:* Remove the pups from the bitch and feed them by hand for twenty-four to forty-eight hours while the bitch receives treatment. Then put the pups back with the bitch. Feed only cold 10 percent glucose until the bloating has disappeared.

## Neonatal Diarrhea

The normal stool of newborn puppies should be brown to yellowish brown and fairly well formed. Under normal circumstances you may never see it because the bitch will clean it away from the pups as it is eliminated. A common problem is a loose, foamy yellow to yellowish green stool. Its presence will often be made known by soiling of the bedding. The cause is usually overfeeding and overstimulation of the digestive system. The digestive enzymes, mainly from the pancreas, are partially depleted, and the liver is overactive and excessive bile is released into the gut, which accounts for the yellowish green color. The pH is lowered (more acid) because of lack of buffering action of the digestive enzymes, and the bowel becomes hyperactive.

The problem may be seen in normal litters that are fat and healthy, presumably when the bitch produces an abundance of milk. Orphan puppies or puppies on supplemental feeding may develop the problem if they are overfed. A change in diet, such as to cow's or goat's milk, could also cause the problem because of the different amounts of fat and lactose.

The problem if left unchecked may proceed in stages from the yellow-green stool to grey to white (the color of milk) because of continued depletion of digestive enzymes. Finally, digestion may be totally halted—a life-threatening situation.

Treatment involves feeding a diet that can be absorbed without digestion until the gut has a chance to repair itself. The elemental diet for premature puppies or glucose in water are useful. Various preparations of electrolytes for babies or farm animals, or Gatorade® may also be useful. Milk of Magnesia® (a few drops for an eight-ounce pup every one to three hours for three to four times) is helpful. It is a laxative and might not seem to be a logical choice to treat a diarrhea, but the benefit comes from the fact that it is also an antacid. The low (acid) pH is one factor that aggravates the situation. Beware that overdosage will act as a laxative and aggravate the diarrhea with possible disastrous consequences. Other antacid preparations could also be used.

## Puppies that Regurgitate Milk

Before fourteen to fifteen days of age, the vagus nerve, which regulates several internal functions, including heart rate, is poorly developed and not fully functional. The tone of the muscle wall of the esophagus depends on vagal stimulation, and some puppies will tend to spit up milk because of an immature, weak esophagus. The problem usually corrects itself with time. Development of vagus nerve activity is also partially responsible for the slower heart rate seen as the puppy matures.

## Puppies with Breathing Problems

There are some puppies whose lungs do not expand fully because they are a little weak at the time of birth. You may see labored breathing or hear fluid or congestion in the chest when they breathe. One technique that can be useful to alleviate this problem is to hold the puppy by its rear legs, head down, at a 45-degree angle. The pup's rear feet could also be taped to a flat object and this held at a 45-degree angle. The pup will be in distress and will cry, thus helping to expand its lungs. The head-down position will also aid in drainage of excess fluid from the lungs. Ideally, this should be done while giving oxygen to prevent true respiratory distress in pups whose lung capacity is still marginal.

## Hemorrhagic Problems

In some kennel situations, hemorrhages, which appear as subcutaneous bruising or nosebleeds, or hemorrhages of the tongue and lips, occur because of vitamin K deficiency. The problem results because of improper storage of dog food under high temperatures, or using food stored too long. Vitamin K is destroyed in the food and hemorrhaging can result. Treatment is injection of vitamin K to the bitches during the last half of pregnancy, as well as to the puppies. The prevention is to buy only fresh dog food from a supplier with a good turnover of its supplies.

## Bacterial Skin Infection

**Juvenile Pyoderma.** *Symptoms*: Crusts, pustules, scabs forming around head and neck, possibly because the bitch does not clean the head as thoroughly as the rear sections of the puppies. *Prevention*: Cleanliness. *Treatment*: Bathe in Phisohex®, Betadine®, antibiotics.

## Neonatal Conjunctivitis

This is an infection of the eyes that occurs before the eyelids are open. Pus may drain from the corner of the eyes or accumulate behind the eyelids and cause an apparent swelling. Treatment includes opening the eyelids, cleaning the eyes, and treating with appropriate antibiotics. The cause may be poor sanitation or infection acquired during whelping, or it may be secondary to another infection.

### Flat Puppy Syndrome

This is a situation in which the pups do not get up on their legs and walk at the appropriate time. The chest cavity may be flat or sunken in. The cause can be simply management—bedding that is too slick for good traction. Nutritional problems are also a possibility, such as excess vitamin D. In some cases, there is a familial tendency which suggests a genetic cause. Affected puppies are sometimes referred to as "swimmers" or spraddle-legged pups. Most puppies will be helped by taping the legs with hobbles to hold them up under the body. Full recovery can be expected in most cases.

# Puppies That Die

I would encourage everyone who has a puppy die, whether at birth or later, to have a post mortem examination (necropsy) done by a veterinarian. In doing so, you will learn much about the weaknesses that may exist in your line, about possible weaknesses in your management routine, and about the general nutritional status of the puppies. It is easy to do, and the cost should be minimal compared with the knowledge that can be gained. Internal birth defects, infections (such as peritonitis), trauma, or physiologic immaturity can be distinguished, whereas the information will be lost forever if you do not look.

The ratio of liver weight to brain weight is important to note at necropsy. Normal pups have a liver that weighs two to four times as much as the brain. When the liver is less than two times as heavy as the brain, it is probably because the puppy was poorly nourished prenatally, and the puppy had a poor chance of survival under any circumstances.

# Internal Parasites

Control of internal parasites, especially ascarids (large roundworms) and hookworms, is an important aspect of successfully rearing healthy puppies. It is generally recommended that the bitch be checked for parasites before breeding and treated as necessary. The purpose of such treatment is to eliminate the stress of being parasitized while pregnant and lactating. In spite of such treatment, a bitch's puppies will be infected with ascarids and/or hookworms if she has been infected with them *at any time* in her life. The life cycle of these parasites makes it nearly impossible to avoid producing infected puppies.

When a dog, either puppy or adult, male or female, is infected with the larval stages of ascarids or hookworms, some of the parasites develop into adults in the intestines, but many of them remain as immature larvae and are encysted indefinitely in various body tissues (muscle, liver and others). The number of larvae in "tissue storage" varies depending on the dog's exposure and immune status. In the case of a male, the tissue-stored larvae never develop further. In the case of a female which is bred, the larvae are activated during the last trimester of pregnancy. They enter the bloodstream, cross the placenta, and infect the unborn fetuses. A bitch previously infected with ascarids or hookworms can produce infected puppies during several pregnancies after her last exposure. By the time the puppies are three to four weeks old, there are adult worms in their intestines. In the case of heavy infections, especially of hookworms, young puppies can be severely debilitated by the time they are two to four weeks old.

Hookworms can also infect newborn pups through the bitch's milk. This does not occur with ascarids.

Management of hookworm and ascarid infections requires early deworming, especially in kennels and areas of the country where these parasites are a problem. Deworming at two to four weeks of age is recommended. Your veterinarian can recommend effective and safe medications to treat these young puppies.

Other parasites, including tapeworm, whipworm, coccidia and giardia are acquired after birth. Good sanitation, early detection and early treatment are important in managing these parasites.

Chapter Nineteen

# WHEN THINGS GO WRONG: REPRODUCTIVE PROBLEMS IN THE BITCH

## Hormonal Abnormalities

Malfunction of the hypothalamic-pituitary hormone system can lead to various problems in reproduction. Unfortunately, it is difficult to pinpoint a specific malfunction. The hypothalamic releasing factors and pituitary trophic hormones are not easy to measure. Their level in the circulating blood may not be especially meaningful anyway; it is the level in the target organ that counts, and there are mechanisms for concentrating the hormones where they act. Serum proteins which bind and carry the hormones are also involved. Failure to cycle, abnormal cycles (too short, too long, split heats, failure to ovulate), and failure to reach puberty could possibly be caused by a hormonal aberration at the highest level.

Hypothyroidism (low levels of thyroid hormones) is a common cause of infertility. Approximately 50 percent of hypothyroid dogs will show reproductive failure. Failure to cycle and/or excessive intervals between heats will generally be seen. Decreased intensity of the heat, prolonged estrual bleeding, or early abortion may also be observed. Other signs of hypothyroidism will probably be observed, such as poor haircoat, dry skin, weight gain, lethargy, and intolerance to cold temperatures. Hypothyroidism is usually a treatable disease. Supplementation of the hormone is routinely helpful; however, the trait follows family lines, suggesting a genetic basis. Hypothyroid dogs and bitches should not be bred unless the problem is known to be acquired—not if you really care about the soundness of the breed.

Hyperadrenalcorticism (high levels of adrenal cortical hormones) will lead to infertility problems in 80 percent of its victims. The adrenal cortex produces cortisol, aldosterone, and male and female sex steroids, and in this condition the levels are abnormally elevated. Prolonged intervals between heats, failure to conceive or to ovulate, or enlargement of the clitoris may be seen. Correcting the adrenal gland problem often does not correct the infertility problems because of degeneration of the ovaries.

Fertility may also be impaired by hypoadrenalcorticism (low levels of adrenal cortical hormones), diabetes mellitus (insulin deficiency that leads to elevated blood glucose), liver cirrhosis (lack of carrier proteins for the gonadal hormones), and by tumors that secrete sex hormones in abnormal quantities.

## Silent Heats

What is generally meant by this term is a situation in which a bitch is in heat, but the signs are not apparent. It is not an abnormality but simply a relative lack of vulvar swelling and colored discharge. Close observation will usually enable such a bitch to be detected in heat, and vaginal cytology will show that the cycle is normal. Breeding at the proper time, determined either by the smears or by her behavior, will be successful.

## Short Cycles

The minimum length of a complete estrous cycle that could be considered normal would be five to five and one-half months. It has been observed that the uterus requires 130 to 150 days following each ovulation to complete its involution and be prepared for the next heat. Some bitches, though, will cycle every three to four months, and many of them are infertile. The reason for this has never really been defined, but it is presumed that in these short-cycling individuals, the endometrium never regenerates fully and implantation is unsuccessful. The only therapy available (and sometimes it is successful) is to keep the bitch out of heat for six months by giving her mibolerone, an androgenic hormone that blocks the pituitary gonadotrophic hormones.

## Long Cycles

The most common cause of excessive intervals between heats is hypothyroidism. Bitches with long cycles (more than fifteen months) that are not hypothyroid usually remain a mystery. Little can be done to help these individuals, whose problem may be abnormal hypothalamic-pituitary function.

# Failure to Cycle

The age at first cycle (puberty) is genetically determined but may be influenced by external factors, including nutrition. It would not be considered truly abnormal to see a first heat as late as two years of age. Beyond that age, the chances of the bitch being normal are slight. Attempts to induce heat are almost always unsuccessful and are not recommended.

A cause of failure to cycle that must be considered in some instances is previous ovariohysterectomy (spay). Hypothyroidism, adrenal hormone abnormalities, or intersex conditions (pseudohermaphroditism or hermaphroditism) may also be involved. A pseudohermaphrodite that shows no external features of its condition, such as an enlarged clitoris, could be checked for its sex chromosome composition (Karyotyping).

# Persistent Heat

This condition is caused by excessive prolonged levels of estrogens. The source of the excess hormone must be determined before any treatment or therapy can be attempted. Ovarian tumors that secrete estrogen, excess administration of estrogenic compounds, and follicular cysts are possible sources. It is important to correct the problem because prolonged elevated estrogen levels are toxic to the bitch and can lead to a potentially fatal anemia. In the case of follicular cysts, surgical rupture of the cysts usually does not correct the condition, and attempts to rupture them with LH injections are also not often curative. Good results have been reported using gonadotropin releasing hormone. Ovariectomy (removal of the ovaries) is often needed.

# Vaginitis

The vagina is not normally a sterile chamber. It harbors a variety of species of bacteria which under normal circumstances are harmless. When pathogenic bacteria invade the vagina, or when the bitch's immune defenses are decreased for any reason, a vaginal infection can result. Studies of the bacterial growth in normal bitches and in bitches suffering a vaginal infection have shown several species of bacteria that may be present in either case. It is not known how bacteria may be transferred between individuals, but individuals living in close association often will have the same vaginal bacteria. The mere presence of bacteria is not the same as the existence of inflammation. Because of this fact, the practice of routinely culturing bitches before breeding or before acceptance to a stud dog is not logical. A vaginal smear is just as informative as well as easier and more economical. Any bitches that show signs of infection and inflammatory response on a smear should be cultured to identify the probable cause and to determine an appropriate antibiotic for treatment.

Signs of vaginitis include abnormal discharge, usually thick and creamy, which indicates the presence of large numbers of pus cells. The color may vary from whitish to gray or cream, tan, brown, red, or green, depending on the stage of the estrous cycle and the organism involved. The vaginal mucosa will be reddened, possibly thickened, and inflamed. Irritation resulting from licking and reddening of the vulva may also be seen. Sour or putrid odor may be obvious. Treatment (by your veterinarian) involves use of vaginal infusions and systemic administration of appropriate antibiotics.

Vaginitis is often seen in pre-pubertal bitch puppies. A bit of sticky, creamy discharge may be seen around the vulva. In most cases, the problem is self-limiting and will disappear with the first estrous cycle. Treatment with antibiotics may be needed in more severe cases.

**Fig. 19.1 Juvenile Vaginitis** *is a common problem in prepubertal bitches, but seldom requires treatment. It generally disappears after the first estrous cycle or after the bitch puppy is spayed.*

Vaginitis may be related to, although may not necessarily be the cause of, infertility. A persistent vaginal infection during estrus, when the cervix is open, may lead to ascending infection of the uterus (endometritis).

## Vaginal Hyperplasia and Polyps

Some older bitches, following the hormonal stimulation of repeated cycles, experience an excessive proliferation of the vaginal lining. The excessive tissue may protrude through the vulva and will generally recede when the current heat is finished. In some individuals, one or several polyps may develop in the vaginal mucosa. These are rounded outgrowths of the mucosa, connected to the vaginal wall by a thin stalk. Small polyps may never be noticed, but larger ones, or ones that originate far caudally in the vagina, may extend through the vulva. It is alarming, to say the least, to suddenly see a round pink mass, like a large egg in some cases, protrude from the vulva. The polyps are benign and can be removed surgically. Only ovariohysterectomy will prevent their reoccurrence.

## Subinvolution of Placental Sites

Normal placental sites in the uterus will shrink and stop bleeding quickly after delivery. In a few cases, this does not happen, and the reason is unknown. Certain placental sites will remain thickened and will continue to bleed. It is usually only a single site within the uterus, but any number may be involved. Fresh red blood will be discharged and is the only sign of the problem. This color must be distinguished from the darker old blood color that is normally seen after whelping. The problem tends to be self-limiting. Various treatments have been attempted, including administration of oxytocin and ergonovine, which cause the uterine muscular tissue to contract. These treatments are frequently unsuccessful, presumably because the abnormality involves the endometrium, not the muscular portion of the uterus. Subinvolution does not seem to interfere with fertility. It may persist for two to three months following whelping but tends to resolve before the next estrous cycle. In severe cases, hysterectomy may be required.

## Postpartum Endometritis

Endometritis is an acute infection/inflammation of the endometrium. Pathogenic bacteria may gain access through the cervix during whelping and lead to the problem. Signs include abnormal vaginal discharge, odor, depression, decreased milk production or abnormal milk, fever, and refusal to eat. Treatment must be aggressive and be started as early as possible. Retained placenta does not cause the problem but may aggravate it by serving as a source of nutrients for the invading organisms. Infusions and systemic antibiotics are required for treatment. Prostaglandin F2$\alpha$ has been used to assist in emptying the uterus if the cervix is open.

## Puerperal Tetany (Eclampsia)

This is a disorder of calcium metabolism, the precise cause of which is unknown. It is seen primarily in small-breed bitches that are about to whelp or have recently whelped and are nursing fat, healthy puppies. The problem is a true emergency, and prompt treatment by a veterinarian must be given to save the life of the bitch. Signs include restlessness, salivation, signs of discomfort, incoordination, stumbling, stiffening of the muscles, and finally convulsions. Treatment is intravenous administration of calcium gluconate, with oral calcium supplementation usually given as a follow-up. Prevention is unknown, although it is generally believed that excessive calcium supplementation during pregnancy may predispose to the problem. Calcium supplements artificially elevate serum calcium levels, and by negative feedback the parathyroid glands, which regulate body calcium levels, become relatively inactive. Later, when body calcium must be mobilized to provide the amounts needed in milk production, the parathyroid glands are unable to respond appropriately, and serum calcium drops to dangerous levels.

## Pyometra

This term refers to pus in the uterus, and pyometra is a serious, potentially fatal uterine disease. The disease process begins with cystic endometrial hyperplasia (proliferative and cystic changes in the endometrium stimulated by hormones, primarily progesterone). Cystic endometrial hyperplasia is an exaggerated response to progesterone and occurs during diestrus. The cystic changes can be produced experimentally by the administration of progesterone, but just why the changes occur naturally is unknown. The progesterone levels are the same in normal bitches and those with pyometra. Add to the situation of cystic hyperplasia some acute inflammation, complicated usually by the presence of bacteria, and you have a severely diseased uterus, enlarged, inflamed, and filled with pus. If the cervix is closed, the symptoms may be more severe, and if the cervix is open, there may be a copious amount of putrid discharge, but the bitch will not be as ill. Signs include sluggishness, refusal to eat, vomiting, distended abdomen, diarrhea, enlarged or flaccid vulva, excessive drinking of water accompanied by urination of large volumes of urine, often urinating during the night, and discharge from the vulva if the cervix is open. The bitch may become quite toxic, and all body systems may be involved if the problem is left untreated. Pyometra usually affects bitches over six years old, although not always; it has occurred after the first heat. It is always seen during diestrus, on the average, five to six weeks following the last estrus. There is no apparent connection, as once believed, with pseudopregnancy, irregular cycles, or the fact that a bitch has never been pregnant.

The usual treatment for pyometra is ovariohysterectomy, and in the case of a closed cervix, this is still the only option available. When the cervix is opened, however, it may be possible to treat the problem medically with prostaglandin F2$\alpha$. This drug causes strong contractions of the uterus and aids in evacuation of the pus. Aggressive supportive care and prolonged treatment with antibiotics are also required, but the treatment can be successful. What a Godsend to breeders to be able to save some bitches for breeding that, not too many years ago, would have had to be spayed.


## Reabsorption, Abortion

It is perfectly natural throughout the animal kingdom for a percentage of pregnancies to fail and for a proportion of conceptuses to fail to survive. Fully a third of human pregnancies fail before full term. In dogs, the same is true but if the entire pregnancy were doomed to failure because of the loss of one embryo or fetus, very few dogs would reproduce successfully. In one study of reproduction in Beagles, it was discovered that of every ten eggs ovulated, nine are accounted for at or shortly after implantation. In other words, 10 percent of potential conceptuses die sometime between ovulation and implantation. Another 10 percent of the conceptuses die between implantation and full term. The embryos that are lost before significant skeletal development are liquefied, along with the placenta and membranes, and reabsorbed. The ones that die later are dehydrated and compacted into a dark brown to blackish mass ("mummy") which is delivered, undetected, at whelping. Late fetal losses result in various amounts of decomposition and compaction ("mummification") depending on the age at death. The fetus plus membranes are delivered along with the normal litter members at whelping. The loss of a member of a litter at any stage should not jeopardize the pregnancy nor the remaining viable fetuses. Reabsorption and mummification are natural processes in any animal that routinely produces litters of offspring.

Reabsorption may result from genetic abnormalities in the embryo/fetus, from abnormalities of the placenta, or from overcrowding in unusually large litters. At times, an entire litter may be reabsorbed, and the same causes are possible, plus the possibility of uterine problems or hormonal abnormalities. Deficient levels of progesterone may result in a failed pregnancy, and supplementation may be successful. As with all such hormonal deficiencies, you must seriously consider whether it is good breeding policy to artificially manipulate defective individuals. In doing so, we may succeed in getting offspring from our prized individual, but at the risk of perpetuating the problem in the next generation.

## Torsion, Prolapse of the Uterus

A uterus enlarged due to pregnancy—heavy and filled with fluid and fetuses—can on rare occasions twist over on itself. When this happens, the blood supply could be acutely and severely compromised, leading to death of fetuses and severe damage to the uterus. In other cases, such an acute crisis might not occur, but the torsion could block the uterus and prevent normal delivery of pups, necessitating a C-section. Discomfort, lethargy, vomiting, abdominal pain, or labor without results could be signs of a uterine torsion and would require prompt veterinary assistance.

Following whelping, another rare complication might be uterine prolapse. In this condition, the uterus turns more or less inside-out, and the everted portion of the uterus bulges through the vagina. The appearance might be identical to that of a large vaginal polyp and should be attended to quickly by a veterinarian. Successful treatment may be possible if started early enough.

## Mastitis

Mastitis refers to infection and/or inflammation of the mammary glands and results from bacterial infection. The route of infection is usually undetermined. Bacteria may invade through the teats or the bloodstream. Painful, reddened, hard, and enlarged mammary glands will be the first signs of trouble. The milk will usually be abnormal, with a creamy appearance indicative of pus, or streaked with blood. If left untreated, the glands may form abscess pockets that can rupture to the outside through the skin, forming open sores. The early stages of mastitis may appear similar to the hard, overfilled glands seen after whelping when the litter is small, or may resemble the condition created when the glands are not being emptied effectively. But in the latter case, the congestion will be quickly relieved if a puppy is placed on the teat to nurse, and the gland will not be reddened and warm. It is good practice to feel every mammary gland twice a day during the early days of lactation and to squeeze a drop of milk from every teat to detect problems in their earliest stages.

## Mammary Tumors

Bitches are quite susceptible to the development of tumors in the mammary glands. It is a preventable disease, however, as the incidence is approaching zero in bitches that are spayed before their first heat. Since spaying is unacceptable for our breeding stock, we must be aware of the problem and be able to detect tumors as early as possible. As with all forms of cancer, early detection is the key to successful cure. The problem may appear as nodules or lumps anywhere along the mammary chain. Many of the tumors are benign, but others are malignant, and nobody can be certain which is which unless a microscopic examination and identification are done by a veterinary pathologist. All masses in the mammary glands should be removed, and in some cases a partial or total mastectomy may be recommended. In some cases the individual may benefit from ovariohysterectomy, as when the tumor is hormone-responsive. In other cases, the spay will not affect the mammary tumor either way. Since mammary cancer can be nearly prevented by early spaying, all purchasers of bitches which are not intended for breeding should be told about this and encouraged to have their pet altered before her first heat.

## Other Tumors

Tumors can arise in any other part of the reproductive system, and the presenting signs are variable. The average age of incidence is nine to ten years. Vaginal tumors can lead to persistent discharge or vaginitis. Internal tumors, involving the uterus or ovaries, are more difficult to detect because of their internal location. Ovarian tumors can produce hormones, including estrogens, with signs referable to the hormone imbalance created.

## False Pregnancy (Pseudopregnancy, Pseudocyesis)

This condition has been mentioned in an earlier chapter. It is a condition in which some signs of pregnancy, including lactation, abdominal distention, and personality changes, occur during diestrus in bitches that are not pregnant. The condition does not relate to previous pregnancies, if any, nor does it interfere with future reproductive performance. Except for the sudden drop in progesterone that occurs shortly before whelping, the levels of progesterone are essentially the same in pregnant, bred but not pregnant, pseudopregnant, and not bred, not pseudopregnant individuals. Generally no treatment is required, but in extreme cases, sex hormones, usually androgens such as testosterone, are given to alleviate the signs.

## Canine Brucellosis

Brucellosis is a disease caused by infection with the bacterium *Brucella canis*. The disease is widespread throughout the United States, although the incidence appears to be low—reportedly 1 to 6 percent of the canine population. The disease is transmitted primarily by the vaginal discharge and mammary secretions of infected bitches and through the semen of males at the time of breeding. Spread through urine or other body secretions is possible though unlikely. Aerosol spread may occur in kennel situations.

Many infected animals appear normal, even though they may have persistent bacteria in their blood. Infected individuals may recover from the infection, but recovery takes two years or longer.

### Signs in the female:

1. Abortion after thirty days post-breeding, most commonly forty-five to fifty-five days.
2. Litters with some dead puppies.
3. Early embryonic death (apparent failure to conceive).
4. Lymph node enlargement.

### Signs in the male:

1. Epididymitis (Inflamed epididymis).
2. Prostatitis (Inflamed prostate).
3. Orchitis (Inflamed testes) followed by atrophy.
4. Pain may lead to licking and dermatitis of the scrotum.
5. Infertility—abnormal sperm (30 to 80 percent early, more than 90 percent late), bent tails, distal droplets, swollen midpieces, reduced motility, detached heads.
6. Lymph node enlargement.

Screening can be done by a rapid slide agglutination test. A negative test is reliable, but a positive test must be rechecked by other methods because false positives are possible. A small blood sample is all that is required.

Tube agglutination tests to detect antibody in infected individuals are needed to confirm infection if the rapid slide test is positive. Cultures are also important to determine if the causative bacteria are present, and are done from blood, semen, vaginal and mammary secretions, lymph nodes, and bone marrow.

Brucellosis is difficult to treat. The causative organisms are apparently sequestered in such a way that they cannot be reached by antibiotics. Treatment is expensive and prolonged, and success is uncertain. Several treatment regimes have reportedly been successful in some cases, and hopefully more reliable methods will be developed in the future.

Brucellosis is considered a serious threat to breeders, not so much because we really expect to see our dogs become infected, but because if they do, the consequences are devastating. Infected animals can never be used in breeding. They should be totally isolated from other animals. In some cases, the only practical way to accomplish this is to destroy the infected animals. The lives of some may be salvaged if they are neutered and treated. In a kennel situation, repeated testing, removal of infected dogs, isolation, and testing of new dogs entering the kennel are needed. Not until all dogs have been tested clear on three consecutive tests at thirty-day intervals can a kennel be considered clear. The problem is not spread only by breeding. Large numbers of organisms are shed in vaginal secretions and contact between dogs at shows and working trials can potentially spread the disease.

Preventive measures include regular testing of breeding stock. Bitches should be tested within thirty days of breeding, and stud dogs every six months. The infection has been reported to have been contracted by humans, and public health officials should be contacted about the possibility in cases of an outbreak.

# Approaches to Infertility in the Bitch

Two categories of bitches need to be considered—those with normal cycles and those with abnormal cycles.

## Normal Cycles

A good record of the bitch's reproductive history is essential in approaching this problem. As a breeder, you should have written records of the reproductive activity of every bitch that has passed through your hands, whether bred or not. There is absolutely no point in trying to remember it all. Precious few of us are blessed with a perfect memory. Records should include the following:

1. Calendar of each cycle of each bitch with daily notes.
2. List of each bitch's heats, dates of onset, interval since last heat, breeding, pregnancy palpation, and whelping dates.
3. Record of pregnancy and whelping, notes on weight gain, temperatures, the progress of whelping.

Daily notes should be brief and are easy to keep. Other records can be kept on file cards or on pages in a loose-leaf notebook. Every time a bitch comes in heat, notes are added to her record, and as each subsequent event takes place, such as breeding or whelping, more notes are made. Writing things down relieves the mind of trying to remember details.

Fully half of the failures to become pregnant in bitches with normal cycles are because of poor breeding management. A look at the history will often reveal that breeding was done too early or too late. The fact that the bitch has cycled normally and has had an apparently normal estrous period indicates that pituitary and ovarian hormone levels are adequate. A thorough physical examination is important, and supporting laboratory tests can be done if indicated, including urinalysis, blood tests, and evaluation of thyroid hormone levels. Progesterone levels during diestrus may be helpful.

The most important potential problems are infection and obstruction of the tubular reproductive organs. Bacterial culture and sensitivity should be done during proestrus when the cervix is open. Even in the presence of a negative vaginal culture, uterine infection can be present and result in infertility. In some cases, antibiotic therapy may be helpful in spite of negative findings. Examination of the vagina for strictures or other abnormalities that may interfere with breeding is important. Examination of the uterus and oviducts is more difficult, but exploratory surgery to examine these organs and possibly to flush through to check for patency of the oviducts and uterus are the most straightforward and conclusive techniques. Contrast radiographic studies are possible but are seldom attempted.

## Abnormal Cycles

A wide range of problems can lead to infertility, and many of these have been mentioned previously. A systematic approach to a bitch with an abnormal cycle is essential. It begins with a thorough history, physical examination and supportive laboratory testing. Usually you must wait for the next estrous cycle to complete a culture and to check regular, preferably daily, vaginal smears. Short or prolonged proestrus, failure to actually develop a normal estrus (as evidenced by failure of the smear to become fully cornified), excessive duration of cornification (which may indicate follicular development with failure to ovulate), and other abnormalities should be detected.

Failure to develop normal estrous behavior may be remedied by artificial insemination at the appropriate time based on vaginal cytology.

Split heat, a condition in which signs of proestrus develop, but the cycle does not continue, followed by an interval of quiescence, and then four to six weeks later a normal proestrus and estrus, can be seen. In the one case I have studied closely, vaginal smears were taken throughout both phases of the split heat. The smear did not com-

plete cornification the first time, although there was obviously follicular activity and elevated estrogens. During the second stage, the
cycle progressed, and breeding occurred normally. Estradiol is elevated in normal bitches four to six weeks before proestrus, and some
bitches may show signs of proestrus at this time.

Hormone assays, primarily for estradiol and progesterone, may
be helpful in pinpointing the nature of the problem in many cases.
Exploratory surgery must remain a possibility, as well, and may provide results more directly and economically than other costly indirect
procedures.

Chapter Twenty

# THE OTHER HALF: REPRODUCTIVE PROBLEMS IN MALES

The male reproductive system is fairly limited in the range of problems that can occur compared with that of the female.

## Developmental Abnormalities

Occasionally, an individual is born that is genetically a male but that has female characteristics, a condition referred to as pseudo-hermaphrodism. In a male pseudohermaphrodite, the gonads are testes, but the duct system and/or the external genitalia may lack full masculine development and possess some female characteristics. They may even look exactly like a female. There is an inherent tendency for the genitalia to develop along the lines of a female, and testosterone from the fetal testes is required for development in the normal male. A pseudohermaphrodite occurs when the fetal testes do not produce testosterone normally. Exposure of a developing fetus to female hormones can also interfere with normal development.

## Primary Infertility

This is a condition in which the male has never been fertile. Often the testes are small, and they may be either abnormally hard or soft. Libido in most cases is not affected. Usually the cause is unknown and cannot be detected. In some individuals, it is undoubtedly an inborn error of development, while in others the cause may be some undetected insult to the testes or epididymis, such as an infection or inflammation. In either case, no treatment or therapy is likely to be of help, although numerous hormone treatments have been attempted.

Causes of primary infertility:

1. Failure of development of parts of the duct system, including absence or occlusion of the epididymis, or ductus deferens.
2. Phimosis, a condition in which the penis cannot be protruded from the prepuce. Usually caused by a persistent fetal structure or too small preputial opening. Can be corrected surgically.
3. Absence of germinal epithelial cells in the seminiferous tubules. "Sertoli-cell-only" syndrome.
4. Errors of metabolism involving the sex hormones, such as testosterone.
5. Klinefelter's syndrome. XXY sex chromosome complement, or other chromosomal abnormalities.
6. Kartagener's syndrome. Defects in the cilia of all the body's ciliated cells, including sperm cells.
7. Sperm cell formation abnormalities.
8. Psychological. Unwillingness to mate, lack of libido.
9. Inguinal or scrotal hernia, with abnormal tissue or organs in the scrotum, creating abnormal bloodflows and/or temperature.
10. Bilateral cryptorchidism.

# Acquired Infertility

The testes are sensitive to environmental influences, and a variety of situations can cause a decline in fertility in a previously fertile individual.

**1. Age:** Atrophy (decrease in size) of testes is a common problem in dogs after ten years of age. The giant breeds experience senile atrophy even earlier, commonly at seven to eight years of age. It becomes important for the owner of an older stud dog to have the dog's semen evaluated regularly, especially if the dog is being offered at public stud. AKC does not recognize litters produced by a male more than twelve years old unless verification is provided of his breeding soundness.

**2. Heat:** Increase of the temperature of the testes to body temperature causes decreased sperm motility and eventually decreased sperm production. The decrease is reversible after a return to normal temperatures, however. Diseases that result in high fever can affect fertility, and it is potentially a problem for individuals that live in the desert Southwest or tropical climates where summer temperatures are routinely well above normal body temperature.

**3. Chemical toxins:** Chemicals that are known to harm the testes, the epididymis, or their function include Cadmium, which can damage Leydig cells and seminiferous tubules. Amphotericin-B, an antibiotic used to treat fungal infections such as coccidioidomycosis, can cause testicular atrophy. Anti-neoplastic (cancer) drugs can inhibit mitotic activity in the testes and thus interfere with spermatogenesis. Many other drugs, especially ones with estrogenic or anti-androgenic activity, will interfere with normal function or cause atrophy. Normal urine, if forced into the epididymis experimentally or during trauma or violent exercise, can cause inflammation of the epididymis and result in infertility.

**4. Torsion:** A complete twist of the testicle results in cutting off of blood supply and will result in loss of capacity to produce sperm in one to two hours. Damage to the testosterone-producing cells may be permanent, so the condition, while rare, should be considered an immediate emergency. Any sudden swelling or discomfort of the dog's scrotum or testes should be brought to your veterinarian's attention immediately.

**5. Environmental stress:** A few individuals when put in new surroundings will develop stress-related changes in the germinal epithelium of the testes. A dog that is taken on an extensive show circuit may experience this problem. The changes are reversible, and a return to normal fertility may be seen after about six months. The mechanism is not known.

**6. Autoimmune reactions:** Spermatozoa are recognized as a foreign substance to the male's body, because normally they are present "outside" the body, away from the immune system. Diseases or trauma can expose the body's immune system to the sperm cells, and they are recognized as any other foreign substance. Antibodies and sensitized white blood cells are produced which then react against the testes. Degeneration and atrophy of the testes and epididymis may be seen in this condition.

**7. Infections:** Any part of the reproductive tract can become infected with various pathogenic bacteria or other microorganisms. Terminology varies with the organ involved:

| Organ | Term Referring to Infection or Inflammation |
|---|---|
| Testis | Orchitis |
| Epididymis | Epididymitis |
| Prostate | Prostatitis |
| Urethra | Urethritis |
| Penis and Prepuce | Balanoposthitis |

Orchitis is an uncommon problem. When it occurs, the testes may become acutely painful and swollen. Any systemic infection that causes a fever can adversely affect spermatogenesis, and the damage could be either temporary or permanent. A testicular biopsy would be needed to determine the extent of damage. Temporary interference with spermatogenesis may not be evident (as abnormal sperm cells in the semen) for as long as eight to ten weeks after the insult. Likewise, recovery might not be detected in the semen for a similar length of time because it takes that long for sperm formation and maturation.

Epididymitis can lead to infertility because of damage to the tubular epithelial cells or scarring of the surrounding connective tissue. Rupture or total occlusion can also result, either of which would result in total sterility if both sides were involved. In the case of rupture or occlusion, sperm may build up in a large mass behind the blockage, forming a cystic pocket called a spermatocoel, or a granulomatous inflammation, a sperm granuloma. Either of these conditions may produce a nodule that is palpable in the epididymis, but neither appears to cause pain.

**Fig. 20.1 A Sperm Granuloma** *may develop following injury to or inflammation of the epididymis and will result in infertility because of blockage.*

The prostate can become extremely painful when infected, and extreme discomfort may be the first sign of trouble. Fever often accompanies prostatitis as well. Other signs might be difficult in defecation or urination if the gland is enlarged, and at times blood may be seen in the urine. The problem requires early and aggressive treatment by a veterinarian. In some cases, abscesses may form in the prostate, even without the usual early symptoms of acute prostatitis. Prostatic abscesses can be difficult to treat and in some cases may require surgery to drain or remove the gland.

Balanoposthitis is a fairly common problem in dogs and in its milder degrees is the source of the creamy discharge seen at the tip of the prepuce. A small amount of such discharge is normal and usually requires no treatment. Simple cleansing of the penis with mild soapy water is all that is needed in more persistent cases. Flushing with mild *Betadyne®* (iodine) solution may be helpful. Trauma to the membranes covering the penis or prepuce can result in severe infections requiring antibiotic therapy, internally, externally, or both.

Urethritis is rarely seen except in connection with severe cases of balanoposthitis, prostatitis, or urinary tract infections. Symptoms include pus or blood in the urine and possibly difficulty in urination if stones are involved.

# Prostatic Problems

The prostate can undergo a slow and progressive enlargement, usually with advancing age, referred to as benign hyperplasia. The condition is usually without detrimental consequences, and the only truly effective treatment is castration. Hormonal therapy can also be attempted. Cystic changes can also be seen in the prostate, giving an appearance similar to that seen with abscesses. Castration and hormone therapy may be helpful in treating this condition as well.

A note concerning the prostate in Scottish Terriers: Scotties have a normal prostate that is much larger than other breeds, as much as four times as large. A Scottie's prostate may seem abnormal to someone examining it who is not aware of this. Also, they tend to produce copious amounts of semen when collected for artificial insemination, presumably because of abundant prostatic fluids.

# Miscellaneous Problems

### Paraphimosis

This is a condition in which the penis, once protruded from the sheath, cannot be normally withdrawn. It can occur during sexual excitement or mating and may be no more than hair wrapped around the shaft of the penis that does not permit the erection to abate. Dry membranes may also contribute to the problem. A small preputial opening may also contribute, constricting the erect penis and not allowing the blood to leave the erectile tissue. In most cases, cold compresses and lubrication will help the swelling of the penis decrease and return to the sheath. In a few cases, the preputial opening must be enlarged surgically. Severe injury to the central nervous system or tumors that interfere with the blood supply may be involved in some rare cases.

## Urethral Prolapse

The urethra can evert, just like any other hollow tubular organ that opens to the outside (e.g. rectum, vagina). When this occurs, a flowerlike, reddened swelling will be seen at the tip of the penis. Lubrication and cold compresses to reduce swelling may be required to return the urethra to its normal position. A veterinarian may use a temporary purse-string suture in the end of the penis or prepuce to reduce the prolapse.

## Tumors

Most cells of the reproductive organs can become involved in formation of tumors, and these can be either benign or malignant. Sertoli cell tumors tend to secrete estrogenic hormones and can cause changes in the skin, haircoat, and physical appearance of the dog (the male feminizing syndrome). Seminomas originate from the germinal epithelial cells of the testis. Both sertoli cell tumors and seminomas are much more common in cryptorchid testes than normal ones, and they occur in older dogs averaging eight to ten years. Interstitial cell or Leydig cell tumors, which are usually benign, also are known to occur but are usually inapparent. They may occur along with other tumors, such as the sertoli cell tumors. The only practical treatment for these testicular tumors would be castration.

Carcinoma of the prostate is rare in dogs, but when it does occur, it tends to be malignant and to metastasize widely.

Transmissible venereal tumor is a unique and rather bizarre tumor seen only in dogs. It is a true neoplasm, but it has a chromosome number (fifty-seven to sixty-two) that is entirely different from normal dog cells (seventy-eight). It is transferred directly from dog to dog during mating by the implantation of tumor cells onto the mucous membrane of the penis or vagina, as the case may be. It is very much like the seed of a mistletoe being planted onto the host tree, taking root and growing as a parasite on the host. The tumor seldom metastasizes, although that has been reported to have happened. Treatment is with surgery, chemicals, and/or radiation, like other tumors. Spontaneous recovery also can occur. The tumor appears as nodules or cauliflower-like outgrowths on the genital organs. The problem is much more common in some sections of the United States than others.

# GLOSSARY OF TERMS

**Acrosome** (ak-ro-sōm). A caplike structure covering the top half of the head of a spermatozoon. Contains enzymes essential for fertilization.

**ACTH, Adrenocorticotropic hormone**. A hormone from the pituitary gland that stimulates the adrenal cortex.

**Adrenal gland** (ad-ré-nal). A glandular organ located in the abdomen adjacent to the kidneys. Secretes several hormones, including cortisone.

**Afterbirth**. The fetal portion of the placenta and the fetal membranes that are discharged after the birth of each puppy.

**Allantoic sac** (ah-lan-to-ic). The inner membrane surrounding the fetus in the uterus.

**Alveoli** (ahl-vé-o-lī). Small sacs lined with secretory cells in the mammary glands.

**Amnionic sac** (am-nēon-ic). The outer membrane surrounding the fetus in the uterus.

**Anatomy** (ah-nat'-o-me). Same as morphology.

**Androgen** (an-dro-jen). Any chemical that has masculinizing activity.

**Anestrus** (an-es-trus). A stage of the estrous cycle during which no reproductive activity is seen. A resting stage.

**Antrum** (an-trum). An opening, as into a cavity or chamber.

**Aplasia** (a-plá-se-ah). Failure of a part or an organ to grow or develop. Lack of the part or organ.

**Aplastic anemia** (ah-plas-tik ah-né-me-ah). A decreased red blood cell and hemoglobin level (anemia) with no tendency to recover by producing new blood cells (aplastic).

**Atrophy** (at'-ro-fe). Degeneration or decrease in size of a tissue or organ.

**Bladder, urinary** (bladder, úri-ner"-e). A hollow, muscular organ in which urine is held temporarily prior to its elimination from the body.

**Blastocyst** (blas-to-sist). An embryo that consists of a hollow ball of cells and an undifferentiated cell mass.

**Brucellosis** (broo-sel-lo-sis). A venereal disease caused by the bacterium *Brucella canis*.

**Bulbus glandis** (bul-bus glan-dis). A portion of the penis. Enlarges during copulation to a round structure commonly called the bulb.

**Bursa** (bur-sah). A pouchlike covering of thin, membranous connective tissue that surrounds each ovary and includes the oviduct.

**Caudal** (kaw-dal). Refers to direction toward the rear or tail end of the body.

**Cervix** (ser-viks). The structure that separates the caudal end of the uterus from the vagina. A part of the uterus.

**Chorion** (ko-re-on). The outermost layer of the embryonic membranes.

**Chromosome** (kro-mo-zōm). A structure in the nucleus of cells that appears during mitosis and that contains the cell's genes. Stains darkly with most stains; roughly rod-shaped.

**Cilia** (sing. cilium) (sil-e-ah). Hairlike outgrowths from the surface of various cells. Their activity tends to sweep fluids and the contents of tubular organs in one direction.

**Cleavage** (klēv-ij). The division of cells in the early embryo, first one cell into two, then four, and so on.

**Cleft palate** (kleft pal-at). A congenital defect in which the two sides of the hard and/or soft palate do not fuse properly.

**Clitoris** (kli-to-ris). Structure in the female that is analagous to the penis. A small structure in the ventral wall of the vestibule.

**Coitus** (ko-i-tus). Same as copulation.

**Colostrum** (ko-los-trum). The first milk secreted after parturition.

**Conceive**. The process of becoming pregnant. Successful union of male and female reproductive cells.

**Conception** (kon-sep-shun). The same as fertilization. The formation of a new individual by the union of male and female reproductive cells.

**Conceptus** (kon-sep-tus). The whole product of conception at any stage of development, from fertilization to whelping.

**Congenital** (kon-jen-i-tal). A problem or defect present since birth. No cause for the problem, such as heredity, is implied.

**Conjunctivitis** (kon-junk-ti-vi-tis). Infection or inflammation of the inner surfaces of the eyelids and/or outer surfaces of the eyeball.

**Copulation** (kop-u-la-shun). The act of mating, or physical union of the male and female, to achieve insemination.

**Cornified** (kor-ni-fi d). Refers to a vaginal smear in which nearly 100 percent of the epithelial cells are fully mature, superficial cells. Also refers to cells that contain keratin.

**Corpus luteum** (pl. corpora lutea) (kor-pus lu-te-um). A glandular organ in the ovaries that develops after ovulation and produces progesterone.

**Cortex** (kor-teks). The outer layer of various glandular organs.

**Corticosteroids** (kor"-ti-ko-ster-oids). Hormones secreted from the adrenal gland cortex. Cortisone is the primary one.

**Cranial** (kra-ne-al). Refers to the direction toward the front or head end of the body.

**Cryptorchid** (krip-tor-kid). Refers to the presence of one or both testes in an abnormal location, in the flank, inguinal canal, or abdomen.

**Culture** (kul-tur). The process of growing bacteria or viruses on artificial growth medium for the purpose of identifying the causative agent of infection.

**Cytology** (si-tol-o-je). The study of cells.

**Cytoplasm** (si-to-plazm"). The substance of a cell, excluding the nucleus.

**Dewclaw** (du-klaw). The small rudimentary digit, homologous to the human thumb, on the dog's foot. Present on the front in all dogs, on the rear in some.

**Diapedesis** (di"-ah-pe-de-sis). The process of migration of blood cells through capillary walls and out into surrounding tissue without disruption of the vessel wall.

**Diestrus** (di-es-trus). A stage of the estrous cycle during which progesterone is the dominant hormone. Follows estrus. Lasts approximately sixty days in the bitch.

**Digital** (dij-i-tal). Refers to the use of the fingers, or digits.

**Diploid** (dip-loid). Refers to a cell that has both members of each chromosome pair normally present for the species. 2N chromosome number.

**Distal** (dis-tal). Refers to orientation away from the point of reference, or away from the center.

**Diuretic** (di"-u-ret-ik). A substance or drug that causes elimination of body fluid through the kidneys.

**DNA, Deoxyribonucleic acid**. The chemical found in the nucleus of all cells that contains its genetic information.

**Dorsal** (dor-sal). Refers to direction toward the back of the body, or in a dog standing naturally, toward the top.

**Ductus deferens** (duk-tus def-er-enz). The tubular outflow organ from the epididymis to the prostate.

**Dystocia** (dis-to-se-ah). Difficult or abnormal parturition.

**Efferent ductules** (ef-er-ent dukt-ul). Small tubules that carry sperm from the testes to the head of the epididymis.

**Ejaculation** (e-jak"-u-la-shun). Expulsion of semen through the male's duct system and urethra to the outside.

**Embryo** (em'-brē-o). Any conceptus after the first cleavage division until the major organ systems and body type have been formed.

**Endometritis** (en''-do-me-tri'-tis). An infection or inflammation of the endometrial layer of the uterus.

**Endometrium** (en-do-me'-trē-um). The inner glandular lining of the uterus.

**Enzyme** (en'-zī m). A chemical substance in cells that serves to regulate other chemical reactions.

**Epididymis** (ep''-i-did'-i-mis). Tubular organ through which sperm pass after leaving the testis. Maturation of sperm occurs here.

**Epithelium** (ep''-i-the'-le-um). The lining tissue of various organs. Can be on the outside or inside, depending on the organ.

**Erection** (e-rek'-shun). The enlargement of the penis due to engorgement of specialized tissue with blood.

**Estradiol** (es''-trah-di'-ol). The female sex hormone that is the most important in the bitch. Secreted by the ovaries.

**Estrogen** (es'-tro-jen). General term referring to the female sex hormones secreted by the ovaries which stimulate signs of estrus.

**Estrous** (es'-trus). Adjective that refers to estrus.

**Estrus** (es'-trus). A stage of the female reproductive cycle during which she is receptive to mating.

**Evisceration** (e-vis''-er-a'-shun). The removal of the organs of the abdominal cavity to the outside because of a break or defect in the body wall.

**Exfoliation** (eks''-fo-le-a'-shun). The loss of cells from the surface of an organ or structure lined by a multi-layered epithelium.

**Fertilization** (fer'-ti-li-za'-shun). The process of union of male and female gametes to form a zygote. Entrance of the spermatozoon into the oocyte and union of their nuclei.

**Fetus** (fe'-tus). An unborn conceptus during the later stages of pregnancy. The species is readily identifiable, and major organ systems have differentiated.

**Fimbria** (fim'-brē-ah). The fingerlike ends of the oviduct adjacent the ovary.

**Flaccid** (flak'-sid). A term used to describe organs or tissues as soft, lacking tone.

**Flagging**. Response of a bitch in estrus. Elevation of the tail and holding it aside, exposing the vulva.

**Follicle** (fol'-li-kl). The oocyte in the ovary and the cells that surround and nurture it.

**FSH, Follicle-stimulating hormone**. A pituitary hormone with action on target cells in the ovaries and testes.

**Gamete** (gam'-ēt). A reproductive cell—oocyte in the female and spermatozoon in the male. Union of male and female gametes is fertilization.

**Gene** (jē n). That part of the cell's genetic material which controls the expression of one physical or chemical trait. Arranged into chromosomes.

**Gestation** (jes-tá-shun). The length of pregnancy; the time from breeding until whelping.

**GnRH, Gonadotropin-releasing hormone**. A hormone from the hypothalamus which influences the pituitary gland to release FSH and LH.

**Gonad** (gōń-ad). The primary sex organ. Testis in the male, ovary in the female.

**Graafian follicle** (graf-e-an). A mature follicle in the ovary filled with fluid, seen shortly prior to ovulation

**Gubernaculum** (gu″-ber-nak-u-lum). A structure that is a part of the caudal ligamentous attachments of the testes. Important in the descent of the testes.

**Haploid** (hap-loid). Refers to a cell that has a single chromosome of each pair normally present for the species. The N chromosome number.

**Hematoma** (hem″-ah-tó-mah). A mass of blood that has escaped from blood vessels and is trapped within tissue.

**Hemoglobin** (he″-mo-gló-bin). The chemical substance present in red blood cells which carries oxygen.

**Hemostat** (he-mo-stat). A surgical instrument that is normally used to clamp vessels to control bleeding.

**Heredity** (he-rcd-i-te). The process of passing traits from one generation to the next through genetic information contained in genes and chromosomes.

**Hermaphrodite** (her-maf-ro-dīt). An animal that has one testis and one ovary. An intersex.

**Hormone** (hoŕ-mōn). A chemical substance produced by cells and released into the bloodstream and having action on target cells in some other part of the body.

**Hyperadrenocorticism** (hi″-per-ad-re″-no-koŕ-ti-sizm). Excessively high levels of hormones from the adrenal gland cortex, or lack of normal daily fluctuation of their levels.

**Hypothalamus** (hi″-po-thal-ah-mus). A portion of the brain adjacent to the pituitary gland. It produces chemical substances that affect secretion of pituitary hormones.

**Hypothyroid** (hi″-po-thi-roid). Abnormally low levels of thyroid hormone. Thyroid hormone helps regulate the metabolism of nearly all body cells.

**Hypoxia** (hi-poḱ-se-ah). Abnormally low levels of oxygen in the body.

**Hysterectomy** (his″-ter-eḱ-to-me). Surgical removal of the uterus.

**ICSH, Interstitial cell-stimulating hormone**. Equivalent to LH. Term sometimes used for LH in males.

**Implantation** (im"-plan-ta'-shun). The process of attachment of the embryo to the endometrium prior to development of the placenta.

**Inguinal canal** (ing'-gwi-nal). An opening in the caudal abdomen through which the testes move in their descent to the scrotum.

**Inguinal ring** (ing'-gwi-nal). The external and internal openings of the inguinal canal.

**Insemination** (in-sem"-i-na'-shun). The process of placing semen into the cranial vagina. May occur during copulation or artificially.

**Intromission** (in"-tro-mish'-un). Process of entrance of the penis into the vagina.

**Involution** (in"-vo-lu'-shun). Regression or decrease in size of the structure or organ after its function has ceased.

**Labia** (la'-be-ah). Folds of skin that form the outer boundary of the female external reproductive organ, the vulva.

**Lactation** (lak-ta'-shun). The process of milk production by the mammary glands.

**Leydig cell** (li'-dig). Secretory cells that lie between the seminiferous tubules. Produce testosterone.

**LH, Luteinizing hormone** (lu'-te-in-i-zing). A pituitary hormone with action on target cells in the ovaries and testes.

**Lochia** (lo'-ke-ah). The discharge from the uterus of fluids and debris following parturition.

**Lock** (lok). Same as tie.

**Locule** (lok'-ul). Same as loculus.

**Loculus** (lok'-u-lus), Pl. loculi (lok'-u-li). The swollen portion of the uterus in which the conceptus is developing.

**Longitudinal** (lon"-ji-tu'-di-nal). Refers to orientation in a lengthwise direction

**Lumen** (lu'-men). The cavity or channel within a tubular organ.

**Lymphatics** (lim-fat'-iks). Vessels and channels that carry lymph, a clear, nearly colorless body fluid.

**Medulla** (me-dul'-lah). The inner layer of various glandular organs.

**Meiosis** (mi-o'-sis). The division of reproductive cells in which the chromosome number is reduced by half.

**Metestrus** (met-es'-trus). A stage of the estrous cycle when a shift from mainly estradiol to progesterone occurs. Cannot be recognized externally. Lasts only a few days.

**Midpiece** (mid'-pes). The part of the spermatozoon that joins the head and tail. Middle piece.

**Misalliance** (mis"-uh-li'-ance). Same as mismating.

**Mismating**. The mating of a dog and bitch that was not intended or desired by the owners.

**Mitosis** (mi-tō'-sis). The division of all body cells in which the chromosome number is maintained and new cells identical to the original are produced.

**Monestrous** (mon-es'-trus). Refers to animals, including dogs, which have long periods of rest between estrous cycles.

**Monorchid** (mon-or'-kid). Refers to the presence of one testis. A true monorchid dog has rarely, if ever, been reported.

**Morphology** (mor-fol'-o-jē). The science that deals with the physical structure of living things. Also, the physical structure of living things.

**Morula** (mor'-u-lah). An embryo that consists of an undifferentiated cluster of cells.

**Mucoid** (mu'-koid). Referring to or having the properties of mucus.

**Mucous** (mu'-kus). Adjective form, refers to the noun mucus. Having the properties of mucus.

**Mucus** (mu'-kus). A substance secreted from glands in certain organs. Usually clear and slippery, serving as lubrication or protection of epithelial surfaces.

**Muscularis** (mus'-ku-la'-ris). The middle muscular layer that makes up the wall of various tubular organs.

**Neutrophil** (nu'-tro-fil). A white blood cell that has a segmented nucleus. One of the body's defense mechanisms to fight infection.

**Nipple** (nip'l). Same as teat.

**Nucleus** (nu'-kle-us). A roughly spherical structure within the cell which stains darkly with common stains and contains the cell's genetic material.

**Omphalocele** (om'-fal-o-sēl). A congenital defect in which the ventral abdominal wall does not fuse properly, allowing abdominal organs to eviscerate.

**Oocyte** (o'-o-sīt). An immature cell in the female germinal cell line. Meiosis occurs in the oocyte stages.

**Oogonium** (o'-o-go'-ne-um). Primitive cell of the female germinal cell line. Divides by mitosis to establish a full population of oocytes.

**Organogenesis** (or'-gah-no-jen'-e-sis). Differentiation and development of the body's organs from undifferentiated cells and tissues.

**Os penis** (os pe'-nis). The small, slender bone that lies within the penis and gives it support during intromission.

**Ovary** (o'-vah-rē), pl. ovaries. The female primary sex organ, or gonad. Located in the abdomen, caudal to the kidneys.

**Oviduct** (o'-vi-dukt). A small, thin, tubular organ that extends from near the ovary to the uterus.

**Ovulation** (ov'-u-la'-shun). The release of a mature oocyte from the follicle during the estrous cycle.

**Ovum** (o'-vum). Same as oocyte, egg.

**Oxytocin** (ok″-se-to-sin). A pituitary hormone that acts to cause uterine contractions during parturition and letdown of milk.

**Palpation** (pal-pa-shun). The process of examination of any part of the body using the sense of touch, usually the hands and fingers.

**Pars longa glandis** (parz long-ah glan-dis). A portion of the penis, the elongated portion cranial to the bulb.

**Parturition** (par″-tu-rish-un). The process of giving birth. In dogs, whelping.

**Pediatrics** (pe″-de-at′-riks). The branch of medicine that deals with the young (puppies or human babies and children).

**Penis** (pe-nis). The male's copulatory organ.

**Perineum** (per″-i-ne-um). The part of the body including the anus and surrounding area to the scrotum or vulva.

**Peristalsis** (per″-i-stal′-sis). Rhythmic segmental contractions of the muscular wall of a tubular organ to move its contents in one direction.

**Peritoneum** (per″-i-to-ne-um). The thin membrane that lines the abdominal cavity and covers the abdominal organs.

**Perivulvar** (per″-i-vul′-var). The area of the body adjacent to the vulva.

**Pipette** (pi-pet′). A slender glass tube used to draw liquids; of various shapes and sizes, usually with a rubber bulb attached to provide suction.

**Pituitary** (pi-tu-i-tar″-e). A gland that lies beneath the brain. Secretes several hormones that regulate other glands and body functions.

**Placenta** (plah-sen-tah). The organ that allows the maternal and fetal circulations to be in close contact for exchange of oxygen, nutrients, and waste products during prenatal development.

**Polyestrous** (pol″-e-es-trus). Refers to animals that have estrous cycles at short intervals, without a rest period between cycles.

**Pregnancy** (preg-nan-se). A state of being in which a female harbors within her reproductive organs developing members of the next generation.

**Prepuce** (pre-pūs). The protective sheath of skin that surrounds the penis.

**Proestrus** (pro-es-trus). A stage of the estrous cycle during which outward signs of heat are seen, and males are attracted, but the bitch will not breed. Precedes estrus.

**Progesterone** (pro-jes-ter-ōn). The hormone produced by corpora lutea of the ovaries. Its primary function is to maintain the reproductive organs during pregnancy.

**Pronucleus** (pro-nu-kle-us). The female or male genetic material in a recently fertilized ovum which has not yet fused with the pronucleus of the opposite sex to produce the nucleus of the zygote.

**Prostate** (pros-tāt). The only accessory sex gland in the dog. Lies in the pelvis at the neck of the urinary bladder.

**Proximal** (prok-si-mal). Refers to orientation near the point of reference or close to the center.

**Pseudocyesis** (su″-do-si-e-sis). Same as pseudopregnancy.

**Pseudohermaphrodite** (su″-do-her-maf′-ro-dī t). An animal with the gonads of one sex and the external sex characteristics of the other. (A male pseudohermaphrodite has testes and appears as a female.)

**Pseudopregnancy** (su″-do-preg-nan-sē). False pregnancy. Expressed to some degree in all diestrous bitches.

**Puberty** (pu-ber-tē). The stage of life at which sexual reproduction is first possible.

**Pyometra** (pi″-o-me-trah). A pathological condition of the uterus in which it is filled with pus.

**Rete testis** (re-te tes-tis). Honeycomb network of small tubules that carry sperm from the seminiferous tubules to the epididymis.

**Scrotum** (skro-tum). The saclike skin that surrounds the testes.

**Semen** (se-men). The fluid released during ejaculation which contains sperm, epididymal fluid, prostatic fluid, and urethral secretions.

**Seminiferous tubule** (se″-mi-nif′-er-us). A long, coiled tubule in the testis, lined with germinal cells and secretory cells and in which spermatozoa are produced.

**Senescence** (se-nes′-ens). The process of aging, and the changes that accompany aging.

**Septicemia** (sep″-ti-se-me-ah). A bacterial infection that has invaded the bloodstream and spread throughout the body.

**Septum** (sep-tum). A sheet of fibrous connective tissue found in various organs. Separates tissues into separate compartments.

**Serosa** (se-ro-sah). The outer layer of the uterus and other tubular organs in the abdomen. A thin layer composed of connective tissue.

**Sertoli cells** (ser-to-lē). Secretory cells that lie in the seminiferous tubules. Secrete hormones and serve as nurse cells for developing spermatozoa.

**Spay** (spā). Surgical sterilization of a female by removal of the ovaries and uterus. Ovariohysterectomy.

**Speculum** (spek-u-lum). A hollow, tubular device that is used to look into body openings, or to provide a clean passage into a body opening for treatment or diagnosis.

**Sperm** (sperm). Same as spermatozoon.

**Spermatocyte** (sper-mah-to-sī t′). Immature cell in the male germinal cell line. Meiosis occurs during the spermatocyte stages.

**Spermatogenesis** (sper″-mah-to-jen-e-sis). The process of formation of spermatozoa.

**Spermatogonium** (sper″-mah-to-go-nē-um). Primitive cell of the male germinal cell line. Divides by mitosis to produce more similar cells.

**Spermatozoon** (sper″-mah-tō-zo-on). Mature cell of the male germinal cell line. Same as sperm.

**Spina bifida** (spi-nah bi-fid-ah). A congenital defect in which the spinal canal does not close properly.

**Squamous** (skwā-mus). Refers to epithelial cells. Flat, scaley, or platelike cells arranged in multiple layers.

**Sternum** (ster-num). The ventral-most part of the chest or thorax.

**Stethoscope** (steth-o-skōp). An instrument that is used to listen to the sound of the heart and lungs through the chest wall.

**Stratified** (strat-i-fi d). Refers to the arrangement of epithelial cells in several to many layers.

**Subinvolution** (sub-in-vō-lu-shun). Incomplete regression or shrinking of the uterus following parturition. May refer to placental attachment sites.

**Teasing**. The process of placing a female in heat and a male together in order to test her behavior, or his, or to collect semen.

**Teat** (tēt). Same as nipple. Raised structure on each mammary gland that provides outflow for milk and access for the newborn to be nourished by suckling.

**Testis** (tes-tis), pl. testes (tes-tēz). The male gonad. Site of production of sex hormones and spermatozoa.

**Testosterone** (tes-tos-ter-ōn). The primary male sex hormone, produced in the testes.

**Theriogenology** (ther-ēō-jen-ol-o-jē). The study of reproduction.

**Tie**. The mechanical holding of the dog's penis in the bitch's vagina during copulation.

**Transverse** (trans-vers). Refers to orientation in the crosswise direction, as opposed to lengthwise.

**Trocarize** (tro-car-ī z). To pierce a cavity with a sharp, hollow instrument to relieve pressure of gas.

**Trophoblast** (trōf-o-blast). The layer of the embryo that attaches to the uterus and forms the fetal part of the placenta.

**Tubal lock** (tu-bal). A functional closure of the oviduct at its junction with the uterus caused by elevated estradiol levels. It prevents ova from moving into the uterus.

**Turgid** (tur-jid). A term used to describe organs and tissues. Firm, inflexible, having good tone.

**Umbilical cord** (um-bil-i-kal). The group of vessels and connective tissue that attaches the fetus to the placenta, taking fetal blood to and from the placenta.

**Urethra** (u-rē-thrah). The tubular organ that connects the urinary bladder to the outside.

**Uterus** (u-ter-us). The ''womb.'' The female reproductive organ in which conceptuses develop during pregnancy.

**Vagina** (vah-jī′-nah). A muscular, tubular organ that extends from the uterus (cervix) to the vestibule. The "birth canal."

**Vaginitis** (vaj″-i-ni′-tis). An infection or inflammation of the vagina.

**Vas deferens** (vas def′-er-enz). Same as ductus deferens.

**Ventral** (ven′-tral). Refers to direction toward the front or "stomach" side of the body. In a dog standing, toward the lower side.

**Vestibule** (ves′-ti-bū l). The chamber in the female reproductive tract that lies between the vagina and vulva.

**Viremia** (vi-re′-me-ah). A virus infection that has spread through the bloodstream to involve all parts of the body.

**Vulva** (vul′-vah). The external opening of the urinary and reproductive systems of the bitch.

**Weaning** (wē n′-ing). The process of separating puppies from dependence on their dam for nutrition by suckling. Separation of puppies from their dam so that they can no longer nurse.

**Whelp** (whelp). Parturition in the dog. Whelping: the process of giving birth. Also as a noun (pl): newborn puppies.

**Zygote** (zi′-gōt). The early embryo, or new individual, produced as a result of fertilization of the oocyte by a spermatozoon.

# REFERENCES AND SUGGESTIONS FOR FURTHER READING

## Articles and Reviews

Baba, E., Hata, H., Fukata, T., and Arakawa, A., "Vaginal and Uterine Micro-flora of Adult Dogs," *Am J Vet Research* 44, No. 4 (1983).

Boucher, J.H., Foote, R.H., and Kirk, R.W., "The Evaluation of Semen Quality in the Dog and the Effects of Frequency of Ejaculation," *Cornell Vet* 48, 1958.

Bowen, R.A., Amann, R.P., Froman, D.P, Olar, T.T., and Picket, B.W., "Artificial Insemination with Frozen Semen in the Dog," *Dog World*, January, 1984.

Concannon, P.W., Hansel, W., and Visek, W.J., "The Ovarian Cycle of the Bitch: Plasma Estrogen, LH, and Progesterone," *Biol. Reprod.* 13, P. 112 (1975).

Currier, R.W., Raithel, W.F., Martin, R.J., and Potter, M.E., "Canine Brucellosis," *J Amer Vet Med Assoc* 180, No. 2 (1982).

Curtis, E.M., and Grant R. P., "Masculinization of Female Pups by Progestogens," *J Amer Vet Med Assoc* 144, No. 4 (1964).

Doak, R.H., Hall, E., and Dale, H.E., "Longevity of Spermatozoa in the Reproductive Tract of the Bitch," *J Reprod Fertil* 13, P. 51 (1967).

Dow, C., "The Cystic Hyperplasia-Pyometra Complex in the Bitch," *Vet Record* 70, P. 1102 (1958).

Ghosol, S.K., LaMarche, P.H., Bhanja, P., et al, "Duration of Meiosis and Spermiogenesis in the Dog," *Can J Genet Cytol* 25 (1983).

Hirsch, D.C., and Wiger, N., "The Bacterial Flora of the Normal Canine Vagina Compared with That of Vaginal Exudates," *J Sm Anim Pract* 18, P. 25 (1977).

Holst, P.A., and Phemister, R.D., "The Prenatal Development of the Dog: Preimplantation Events," *Biol Reprod* 5, P. 194 (1971).

**218**

Holst, P.A., and Phemister, R.D., "Onset of Diestrus in the Beagle Bitch: Definition and Significance," *Am J Vet Res* 35, P. 401 (1974).

Holst, P.A., and Phemister, R.D., "Temporal Sequence of Events in the Estrous Cycle of the Bitch," *Am J Vet Res* 36, P. 705 (1975).

Jezyk, P.F., "Metabolic Diseases—An Emerging Area of Veterinary Pediatrics," *Compend on Contin Educ.* 5, No. 12 (1983).

Jochle, W., and Andersen, A.C., "The Estrous Cycle in the Dog: A Review," *Theriogenology* 7, P. 113 (1977).

Johnson, C.A., Bennett, M., Jensen, R.K., and Schirmer, R., "Effect of Combined Antibiotic Therapy on Fertility in Brood Bitches Infected with *Brucella canis*," *J Amer Vet Med Assoc* 180, No. 11 (1982).

Johnston, S.D., "Diagnostic and Therapeutic Approach to Infertility in the Bitch," *J Amer Vet Med Assoc* 176, No. 12 (1980).

Johnston, S.D., Smith, F. O., and Barr, R.I., "Prostaglandin Treatment for Pyometra in the Bitch," *AKC Gazette*, February, 1981.

Johnston, S.D., Smith, F.O., Bailie, N.C., Johnston, G.R., and Feeney, D.A., "Prenatal Indicators of Puppy Vitality at Term," *Compend on Contin Educ.* 5, No. 12 (1983).

Kornblatt, A.N., Adams, R. L., Barthold, S.W., and Cameron, G.A., "Canine Neonatal Deaths Associated with Group B Streptococcal Septicemia," *J Amer Vet Med Assoc* 183, No. 6 (1983).

LaCroix, J.A., "A Caesarian Section (photos)," *AKC Gazette*, 1982.

Larsen, R.E., "Evaluation of Fertility Problems in the Male Dog," in *The Veterinary Clinics of North America* (Philadelphia, PA: W.B. Saunders, November, 1977).

Mellin, T.N., Orazyk, G.P., Hichens, M., and Behrman, H. R., "Serum Profiles of Luteinizing Hormone, Progesterone and Total Estrogens During the Canine Estrous Cycle," *Theriogenology* 5, P. 175 (1976).

Mosier, J.E., "Canine Pediatrics," recorded lectures, distributed by Audio Veterinary Medicine, P.O. Box 2926-D, Pasadena, Calif., 91105, 1979.

Mosier, J.E., "The Effects of the Post-Parturient Bitch on Puppy Health," American Animal Hospital Association, 49th annual meeting, proceedings, 1982.

Mosier, J.E., "The Puppy from Birth to Six weeks," in *The Veterinary Clinics of North America* (Philadelphia, PA: W.B. Saunders, February, 1978).

Nelson, R. W., Feldman, E.C., and Stabenfeldt, G.H., "Treatment of Canine Pyometra and Endometritis with Prostaglandin F2 Alpha," *J Amer Vet Med Assoc* 181, No. 9 (1982).

Nett, T.M., Akbar, A.M., Phemister, R. D., Holst, P.A., Reichert, L.E., Jr., and Niswender, G.D., "Levels of Luteinizing Hormone, Estradiol, and Progesterone in Serum During the Estrous Cycle and Pregnancy in the Beagle Bitch," *Proc Soc Exper Biol Med* 148, P. 134 (1975).

Olson, P.N., Husted, P.W., and Nett, T.M., "The Management of a Successful Mating Between the Bitch and the Stud Dog," *Kal Kan Forum* (1983), pp. 15-21.

Olson, P.N.S., and Mather, E.C., "Canine Vaginal and Uterine Bacterial Flora," *J Amer Vet Med Assoc* 172, P. 708 (1978).

Osbaldiston, G.W., Nuru, S., and Mosier, J.E., "Vaginal Cytology and Micro-flora of Infertile Bitches," *J Amer Anim Hosp Assoc* 8, P. 93 (1972).

Phemister, R.D., Holst, P.A., and Lee, A.C., "Irradiation of the Canine Con-ceptus: Teratogenic, Lethal and Growth-Retarding Effects," 5th ICLA Symposium, Gustav Fischer Verlag, Stuttgart, 1973.

Phemister, R.D., Holst, P.A., Spano, J.S., and Hopwood, M.L. "Time of Ovulation of the Beagle Bitch," *Biol Reprod* 8, P. 74 (1973).

Picket, B.W. (editor), Proceedings of 1981 canine reproductive management and A.I. short course, Animal Reproduction Lab, Colorado State Univer-sity, Fort Collins, Colorado 80523.

Pineda, M.H., Kainer, R.A., and Faulkner, L.C., "Dorsal Median Postcervical Fold in the Canine Vagina," *Am J Vet Res* 34, No. 12 (1973).

Rosenthal, R.C., "Infertility in the Male Dog," *The Compend on Contin Educ* 5, No 12 (1983).

Seager, S.W.J., and Fletcher, W. S., "Collection, Storage and Insemination of Canine Semen," *Lab Anim Sci* 22, No. 2 (1972).

Senior, D.F., "Infertility in the Cycling Bitch," *The Compend on Contin. Educ* 1, P. 17 (1979).

Shores, A., "Neurologic Examination of the Canine Neonate," *The Compend on Contin Educ* 5, No. 12 (1983).

Smith, M.S., and McDonald, L.E., "Serum Levels of Luteinizing Hormone and Progesterone During the Estrous Cycle, Pseudopregnancy and Pregnancy in the Dog," *Endocrinology* 94, P. 404 (1974).

Sokolowski, J.H., "Reproductive Patterns in the Bitch," in *Veterinary Clinics of North America* (Philadelphia, PA: W.B. Saunders, 1977).

Sokolowski, J. H., Stover, D.G., and VanRavenswaay, F., "Seasonal Incidence of Estrus and Interestrous Interval for Bitches of Seven Breeds," *J Amer Vet Med Assoc* 171, No. 3 (1977).

VanderWeyden, G.C., Taverne, M.A.M., Okkens, A.C., et al., "Intra-uterine Position of Canine Fetuses and Their Sequence of Expulsion at Birth," *J Small Anim Pract* 22 (1981).

Wildt, D.E., Chakraborty, P.K., Panko, W.B., and Seager, S.W.J., "Relation-ship of Reproductive Behavior, Serum Luteinizing Hormone and Time of Ovulation in the Bitch," *Biol Reprod* 18, P. 561 (1978).

Wildt, D.E., Levinson, C.J., and Seager, S.W.J., "Laparoscopic Exposure and Sequential Observation of the Ovary of the Cycling Bitch," *Anat Record* 189, P. 443 (1977).

Zoha, S.J., "Effect of a Two-Stage Antibiotic Treatment Regimen on Dogs Naturally Infected with *Brucella canis*," *J Amer Vet Med Assoc* 180, No. 12 (1982).

# Books

Andersen, A.C., and Simpson, M.E. *The Ovary and Reproductive Cycle of the Dog (Beagle)* (Los Altos, CA: Geron-X, Inc., 1973).

Burns, M., and Fraser, M. *Genetics of the Dog—The Basis of Successful Breeding*, 2nd ed. (    :Oliver and Boyd, Ltd, 1966).

Clark, R.D., and Stainer, J.R., eds., *Medical and Genetic Aspects of Purebred Dogs* (Bonner Springs, KS: Vet Med Publishing Co., 1983).

Concannon, P.W., "Reproductive Physiology and Endocrine Patterns of the Bitch," in *Current Veterinary Therapy, VIII* Philadelphia, PA: W.B. Saunders, 1983).

Erickson, F., Saperstein, G., Leipold, H.W., and McKinley, *Congenital Defects in Dogs—A Special Reference for Practitioners* (    :Veterinary Practice Publishing Co., 1978). Reprinted for Ralston Purina Co.

Evans, H.E., and Christensen, G.C., "Reproduction and Prenatal Development," chapter 2 in *Miller's Anatomy of the Dog*, 2nd ed. (Philadelphia, PA: W.B. Saunders 1979).

Holst, P. A., *The Preimplantation Development of the Dog*. M.S. thesis, Colorado State University, Fort Collins, Co., 1970.

McDonald, L.E., *Veterinary Endocrinology and Reproduction* (Philadelphia, PA: Lea and Febiger, 1971).

Morrow, D.A., ed., *Current Therapy in Theriogenology* (Philadelphia, PA: W.B. Saunders Co., first ed. 1980, second ed. 1984).

Nett, T.M., and Olson, P.N., "Reproductive Physiology of Dogs and Cats," in *The Textbook of Veterinary Internal Medicine II*, S.J. Ettinger (ed.) (Philadelphia, PA: W.B. Saunders 1982).

Smith, H.A., Jones, T.C., and Hunt, R.D. (eds.), *Veterinary Pathology*, ed. 4, Chapter 26, The Genital System (Philadelphia, PA: Lea and Febiger, 1972).

Stabenfeldt, G.H., and Shille, V.M., "Reproduction in the Dog and Cat," in *Reproduction in Domestic Animals*, 3rd ed., H.H. Cole and P.T. Cupps (eds.) (New York: Academic Press, 1977).

Vanderlip, S.L., *The Collie: A Veterinary Reference for the Professional Breeder* Biotechnical Veterinary Consultants, P.O. Box 327, Cardiff by the Sea, CA. 92007, 1984.

# Index

Gestation length, 133-135
GnRH (Gonadotropin Releasing
Hormone), 27, 28, 29, 30, 33
Gubernaculum, 7, 20, 21

Herpes viremia, 181
Hyperadrenalcorticism, 185-186
Hypothalamus, 27, 28, 29, 30, 33
Hypothyroidism, 185

ICSH (Interstitial Cell Stimulating
Hormone), see LH
Infertility,
female, 195-197
male, 110, 199-203
Inguinal ring, canal, 7, 16, 20, 21
Interstitial (Leydig) Cells, 17, 28-29

Lactation, 116-117, 151-155
Leydig Cells, see Interstitial Cells
LH (Luteinizing hormone), 27-30,
33, 34, 37, 109-110
Litter size, 95, 108-110, 133, 135
Long cycles, 186-187

Mammary Glands, 14, 125,
151-155,
tumors, 192-193
Mastitis, 192
Megestrol acetate, 107-108
Meiosis, 18, 36-37, 38
Metestrus, 35, 39
Mibolerone, 108, 110
Milk, 154
acid, 154-155
regurgitating, 178
toxic, 181
Misalliance = Mismating, 63, 95

Newborn puppies,
normal, 146-149, 157-166
orphans, 166-170
premature, 171-172
problems, 175-183
death, 182
Nutrition, 113-118, 167-168, 171

Oocyte, 9, 36, 37-39
Ovaban, see Megestrol acetate
Ovary, 5-10
Oviduct, 7, 8, 11, 33

Ovulation, 6, 10, 11, 29, 33,
34-37, 61
Ovum, see Oocyte
Oxytocin, 30, 145-146, 153

Parasites, 182
Penis, 7, 15-16, 20, 24-25,
mechanism of erection, 25,
in Artificial Insemination, 97-98
problems, 203-204
Peritonitis, 180
Placenta, 131-132, 139-140
Postpartum, 149
Pregnancy, 39, 114-116
detection, 119-122,
signs, 122-126, 152
Prenatal development, 127-131
Prepuce, 15, 20, 24-25
Proestrus, 33-34, 35, 40, 41, 65
Progesterone, 11, 17, 29, 33, 34,
35, 39, 78, 111, 132, 135, 152
Prostaglandins, 111, 190-191
Prostate, 7, 15-16, 20, 23-24
problems, 202-203
Pseudopregnancy, 39, 193
Pyometra, 111, 190-191

Reabsorption, 191
Record keeping, 75-76, 78

Scrotum, 7, 20, 21
Semen, 15, 80-84, 97-104
frozen, 104
Septicemia, 180
Sertoli Cells, 16-17, 27, 28, 29
Sex ratio, 36, 94
Short Cycles, 40, 110, 186
Silent Heats, 186
Sires, multiple, for a litter, 87
Skin problems, 179
Spay incontinence, 106
Spermatozoa (Sperm), 15-19
lifespan, 19
Capacitation, 18-19
Spermatogenesis, 18, 27, 36,
80-84
Subinvolution, 189

## Your Comments are Invited

If you enjoyed *Canine Reproduction,* or if you found it especially helpful, we would like to hear from you. If you would like to comment on some aspect of the book, feel free to do so. Just write:

Editorial Office
Alpine Publications
225 S. Madison Ave.
Loveland, CO 80537

## For a Free Catalog of Alpine Books

or for information on other Alpine Blue Ribbon titles, please write to our Customer Service Department, P. O. Box 7027, Loveland, CO 80537, or call toll free 1-800-777-7257.

## Additional Titles of Interest:

*How to Raise a Puppy You Can Live With*
Rutherford and Neil
This book is a "must" for every breeder, as well as for new puppy owners!

*Canine Hip Dysplasia*
Fred L. Lanting

*The Health of Your Dog*
Bower and Youngs

and various single breed books.